MANPOWER AND HUMAN RESOURCES STUDIES

NO. 3

MANPOWER IN HOMEBUILDING:
A Preliminary Analysis

by

HOWARD G. FOSTER

Associate Professor of Industrial Relations
State University of New York at Buffalo

Published by
INDUSTRIAL RESEARCH UNIT
The Wharton School
University of Pennsylvania

Distributed by
University of Pennsylvania Press
Philadelphia, Pennsylvania 19174

An earlier version of this study was prepared for the Manpower Administration, U.S. Department of Labor, under Grant Number DL 91-36-71-41. Since grantees preparing research under government sponsorship are encouraged to express their own judgment freely, this study does not necessarily represent the Manpower Administration's official opinion or policy. Moreover, the author is solely responsible for the factual accuracy of all material developed in this study.

Library of Congress Catalog Card Number 74-12751
MANUFACTURED IN THE UNITED STATES OF AMERICA
ISBN: 0-8122-9086-0

Foreword

For many years, the Industrial Research Unit has sponsored studies designed to improve our knowledge of America's manpower and its human resources. These studies began with the forming of the Unit in 1921 under the leadership of Professor Joseph Willits and Anne Beginson, and were especially emphasized during the 1950's when the late Gladys L. Palmer served as Director. Since the mid-1960's, our manpower research has been concentrated on the problems of black employment and the results have been published thus far in the thirty-one monographs of the Racial Policies of American Industry series and the seven volumes of the Studies of Negro Employment.

The Industrial Research Unit faculty staff has determined to continue the two series dealing with black employment, but also at this time to direct part of its time and resources to broader manpower themes. Accordingly, we have inaugurated a new series, Manpower and Human Resources Studies. Two existing studies, Gladys Palmer's last great work, *The Reluctant Job Changer,* (1962), and *Educating the Employed Disadvantaged for Upgrading,* by Richard L. Rowan and Herbert R. Northrup (1972) will be henceforth included in this series as Nos. 1 and 2, respectively.

This study, No. 3 in the new series, *Manpower in Homebuilding: A Preliminary Analysis,* is a thorough revision of a report prepared for the Manpower Administration, U.S. Department of Labor, under Grant No. DL 91-36-71-41. It marks the first attempt to examine how employers in this highly competitive, largely nonunion industry segment employ, direct, compensate, train, and otherwise utilize manpower in the light of the fact that no central market governing force, for example, a union hiring hall, exists. Throughout the study a comparison is made with manpower policies in the union sector while at the same time making careful note of the fact that nonunion, or "open shop" construction, extends far beyond the housing segment.

Nearly all previous studies of construction manpower have concentrated their attention exclusively on the unionized sector. *Manpower in Homebuilding* is the first of two studies undertaken by the Industrial Research Unit, and is preliminary to a major work, scheduled for early 1975 publication, which explores the entire open shop construction part of the industry nationwide. *Manpower in Homebuilding* is also "preliminary" in that data collection was concentrated in one market area, although all findings were checked against conditions in other localities.

Whatever limitations that it might have, *Manpower in Homebuilding* is a path-breaking study of a very significant, but largely unstudied sector of the economy. The author, Professor Howard G. Foster, State University of New York at Buffalo, has devoted a considerable portion of his research over the years to construction labor problems, and has published several articles thereon. He is also playing a major role in the aforementioned study of open shop construction.

Many persons and organizations assisted in this study, including numerous contractors and others who gave freely of their time for interviews. Dr. Foster wishes also especially to thank Dr. Howard Rosen, Director, Office of Research and Development, Manpower Administration, for the original grant and for many thoughtful comments; Mrs. Ann L. Emerson and Mrs. Kathleen G. Messina, who edited the manuscript, and Mrs. Messina, who did the index as well; numerous secretaries at the Industrial Research Unit and the State University of New York at Buffalo for much typing and retyping; and Mrs. Margaret E. Doyle, our office manager, for handling various administrative details. Permission to reuse materials originally printed in the author's articles in the *Industrial and Labor Relations Review*, the *Monthly Labor Review*, the *Quarterly Review of Economics and Business*, and the *Labor Law Journal* was kindly given by the respective editors. The author, of course, is solely responsible for the facts and opinions set forth, and they in no way should be attributed to the Manpower Administration, the State University of New York at Buffalo, or the University of Pennsylvania.

HERBERT R. NORTHRUP, *Director*
Industrial Research Unit
The Wharton School
University of Pennsylvania

Philadelphia
August 1974

TABLE OF CONTENTS

LIST OF TABLES

CHAPTER I

Introduction

In recent years, the operation of labor markets in the construction industry has received considerable scrutiny from both scholars and policy makers concerned with contemporary manpower issues. Both academic and government periodicals are devoting more and more space to the particular problems of the industry.[1] Meetings of professional associations almost invariably include sessions dealing with one or another aspect of construction labor. A comprehensive study of the industry's labor markets and labor relations has recently been published and is the first in many years.[2] Most of this research, moreover, is directed toward important social and economic policy questions: the U.S. Manpower Administration, for example, sponsored more than one dozen major studies on construction labor in the 1971-1972 fiscal year alone.

The reasons for this extensive and growing attention are several. First, and most obvious, the construction industry itself is a large one. In 1972, new construction was valued at over $123 billion, or 10.7 percent of the gross national product that year.[3] Contract construction, which constitutes some 70 percent of all construction activity measured by money value, employs directly about 3.5 million persons on annual average.[4] Since the labor force experiences an inordinate amount of turnover, the industry, therefore, provides the bulk of employment for over

---

[1] For an interesting discussion of the problems experienced in construction, see "Symposium: Industrial Relations in Construction," *Industrial Relations*, Vol. XI (October 1972), pp. 289-411.

[2] Daniel Quinn Mills, *Industrial Relations and Manpower in Construction* (Cambridge: MIT Press, 1972). This study deals primarily with the unionized sector of the construction industry.

[3] U.S. Department of Commerce, *Construction Review*, Vol. XIX, no. 10 (October 1973), p. 7.

[4] *Ibid.*, p. 52.

five million workers.[5] Millions more are employed in related industries such as building materials manufacturing and real estate. In addition, about one-fourth of all construction is done for public purposes, which along with that portion of private building subsidized by various levels of government generates a substantial and legitimate social concern for the operations of the industry. Moreover, construction has been the only goods-producing industry whose employment has grown as fast as that of the economy generally.

Construction has also generated widespread public concern because of its central position in a host of pressing and controversial national problems. The industry consistently suffers an unemployment rate double that of the economy as a whole. It has contributed substantially to the inflationary pressures of recent years, so much so that stabilization policies were imposed upon construction long before the general wage-price freeze of August 1971.[6] Its strike record has been demonstrably worse than that of most other industries. It is widely regarded as a foremost bastion of racially restrictive employment practices. Furthermore, the "housing problem" can hardly be addressed without some reference to the manpower and economic problems of the industry.

Finally, the construction labor market itself is unique. Although some of its features are common to other industries, none other exhibits a combination of characteristics that places such a strain on the ability of a market to perform its intended functions. Most construction products are immobile and are fashioned to precise specifications; hence, the work force necessary to complete a project is usually different from the work force on most other projects. Once a structure is completed, work groups must be dismissed and reassembled, with somewhat different components, for another undertaking. The demand for construction is highly unstable, both from year to year and from season to season, so that job opportunities are irregular and somewhat unpredictable. In addition, jobs require on the average a greater degree of skill than in most other industries, so that the need for lead time in training new workers is es-

---

[5] U.S. Department of Labor, *Manpower Report of the President* (Washington, D.C.: Government Printing Office, 1973), p. 184.

[6] Nor was special treatment for construction in this regard unprecedented. See Daniel Quinn Mills, "Construction Wage Stabilization: A Historic Perspective," *Industrial Relations, op. cit.*, pp. 350-365.

pecially acute. All these factors combine to require labor market processes which are extraordinarily flexible and responsive.

THE CONSTRUCTION INDUSTRY

Construction activity may be defined as the combination of various resources—labor, land, machinery, and building material—which results in the fabrication and maintenance of immobile physical structures. The construction *industry*, as conventionally regarded by labor market analysts, is a somewhat more narrow concept, comprising those individuals and organizations engaged in the erection of structures for the ultimate use of others. Thus, although a business or a governmental agency may perform construction work on its own account, such an enterprise would not normally be included in labor market data on the construction industry. The term "contract construction" is used to designate those firms which perform building work for others, and it has been estimated that about 70 percent of all construction put-in-place is done on contract.[7] The term excludes builders, predominantly residential, who erect buildings for inventory and subsequent rental or sale without a prior contract. These firms are called "operative builders" and are classified in government statistics under real estate.

The construction industry may be divided conceptually several ways with the most appropriate method depending on the subject areas under consideration. One approach is to distinguish the construction product according to its nature and use. *Building* construction includes the erection of buildings for residences (homes, apartments, hotels); commercial use (stores, offices, warehouses); industrial use (factories, blast furnaces, refineries); and so forth. *Highway* construction includes the production of streets, sidewalks, and highways, as well as their maintenance. *Heavy* construction includes structures such as bridges, tunnels, dams, pipelines, harbor projects, and the like.

Another way of dividing the industry is according to the functions of the various firms involved. Government data usually distinguish between general contractors and special trade contractors (i.e., subcontractors) with general contractors further divided into those specializing in building and nonbuilding work. Special trades contractors are those which perform certain spe-

---

[7] John R. Cambern, "Profile of the Contract Construction Industry," *Construction Review*, Vol. 10, no. 9 (September 1964), p. 4.

cific functions within the total project, such as electrical work, plumbing and/or heating, painting, and tile setting.

The industry may also be divided according to the ownership of the structure, with the most significant distinction between public and private. Many classes of projects include substantial amounts of both public and private work (schools, hospitals, office buildings, etc.), although most residential and commercial work is private and most highway work, public. Some types of heavy construction are predominantly public (bridges, dams, sewage treatment facilities), whereas others are mostly private (pipelines, telephone, and power lines).

The Residential Sector

Residential construction has always been a major component of the industry. In 1972, the value of residential building constituted over 40 percent of all construction put-in-place, of which nearly two-thirds was in single-family units.[8] Furthermore, the residential sector is particularly volatile, especially at the local market level, exhibiting a marked sensitivity to both seasonal conditions and year-to-year fluctuations in demand. For these reasons alone, one can easily justify focusing research efforts on this segment of construction.

In addition, however, the residential sector of the industry may be said to merit special concern for at least two other important reasons. First, there is the indisputable social significance of the product. As the population continues to grow and become more affluent, as the postwar baby boom continues to reach the home-buying age, and as housing in our central cities continues to decay, the question of how (or whether) the industry can adapt to these growing demands assumes ever greater importance. If any public policy is to be fashioned to facilitate this adjustment, it is obviously necessary to understand how the labor market in the residential construction industry presently operates.

Second, there is a relative absence of unions in the homebuilding industry. Although construction on the whole has been traditionally one of our most heavily unionized industries, there is, nevertheless, a substantial nonunion work force and it is widely recognized that homebuilding constitutes a major portion of this nonunion activity. Although nonresidential construction is

---

[8] U.S. Department of Commerce, *loc. cit.*

by no means completely unionized, homebuilding is more pervasively nonunion than the other segments of the industry. The sparseness of unionization in residential construction is significant because of the union's dominant role in the labor market in the other sectors of the construction industry, primarily with regard to hiring practices, apprenticeship, and wage determination.

For present purposes, the residential sector may be viewed as consisting of two subsectors: the first includes the building and remodeling of single-family homes, duplexes, and small apartment buildings; and the second, high-rise apartments. The distinction is important because the large apartment complexes are frequently constructed with union labor, whereas the market for smaller structures is predominantly nonunion. Since a major purpose here is to analyze labor market operations in the absence of unions, we shall concentrate on the single-family subsector. A more detailed description of the structure of the residential construction industry in Erie County will be offered in the following chapter.

In a previous study of construction labor in upstate New York, the following observation was offered:

> The most neglected area of construction manpower is that in the nonunion sector. And for good reason. A dispersed, localized industry like construction is difficult enough to study even with the centralizing influence of labor unions, to say nothing of the role unions can play in providing information and access to the workers themselves. Nevertheless, the nonunion sector of construction is too large to be ignored.
>
> The possibilities for research in nonunion construction are virtually endless, as are the problems inherent in such research. All the issues raised . . . are potential candidates for investigation in the nonunion context. The impact of seasonality, seasonal labor force adjustments, training, the various dimensions of mobility, hiring practices—in short, every aspect of the labor market in nonunion construction—are virgin territory for the researcher.[9]

Those words, written five years ago, are largely true today. This study, therefore, is an attempt to expand our ever-growing understanding of the construction labor market into the hitherto unexplored area of nonunion homebuilding, and as noted in the Foreword, it is written primarily to precede a more comprehensive study of the open shop segment of the industry, which is now in progress.

---

[9] Howard G. Foster, "Labor Supply in the Construction Industry: A Case Study of Upstate New York," Ph.D. dissertation, Cornell University, 1969, pp. 362-363.

OBJECTIVES

The objectives of this study are both exploratory and analytical. There are a number of questions to be critically examined, but any analytical illumination is at present severely limited by the paucity of empirical information on the homebuilding labor market. Unfortunately, most government data do not distinguish homebuilding construction from the other sectors of the industry, nor nonunion construction from the union sector. Furthermore, almost all previous academic research has been limited to issues in which the building trades unions play a dominant role. Thus a significant overall purpose here will be to utilize available meager information and to generate some primary data upon which meaningful conclusions can be based.

The basic function of a labor market is to provide a mechanism whereby the supply of workers can be adjusted to an exogenously determined change in demand, thereby allocating human resources according to their most productive potential. This mechanism includes a number of elements: to attract workers when jobs are plentiful and to dismiss them when jobs are scarce; to help meet seasonal needs as well as replacement and growth needs; to transmit messages through wage rates in response to changes in both aggregate demand and the composition of demand; and to facilitate the individual matching of available jobs to unemployed workers.

In turn, elements of the labor market mechanism suggest the four broad areas which will be examined in this study. One primary factor to be considered is the seasonal instability to which the construction industry is subject. The market must somehow provide a peak-period work force much larger than can be sustained throughout the year. Here, the magnitude of this problem and the market's adjustment to it will be explored.

A second factor to be analyzed is the replacement and growth needs of the industry. Problems encountered here arise when workers must be found to replace those leaving the industry through death, retirement, and mobility, as well as to provide for secular growth in demand. Since a large proportion of the work force is skilled, inquiry must especially be made into the methods by which new workers acquire the necessary training.

The third factor, matching supply to demand, presents a problem to the nonunion homebuilding sector since there is no equivalent mechanism to the union hiring hall which is so prominent in the union sector of the construction industry. We propose here to describe and to evaluate the job matching process in homebuilding, comparing it to the prevailing system in the union sector.

The fourth area to be considered is wages and benefits. It is easily demonstrated that basic wages in homebuilding are substantially lower than those in commercial construction, therefore, questions pertaining to the extent, the causes, and the implications of the development and maintenance of this differential should be studied.

Finally, no study of contemporary construction labor markets can ignore the issue of employment discrimination. For more than a decade, public attention has been directed toward restrictive practices in construction which are said to bar the entrance of minorities to high paying jobs in the industry. Unfortunately, none of the field research which forms the basis for this book dealt directly with the problem of racial barriers; however, the implications of labor market behavior and patterns on the prospects for improving employment opportunities for minorities in the housing sector will be discussed. Further studies by the Industrial Research Unit will explore this question in greater detail for the entire industry.[10]

SCOPE AND METHODOLOGY

The primary source of information for this study is based on a detailed survey of residential employers in Erie County, New York. Erie County is composed of three cities and twenty-five townships with a total 1970 population of about 1.1 million,[11] roughly the same as in 1960. Although nearly one-half of the county's population resides in the city of Buffalo, virtually all new residential construction takes place outside the city limits.

As with most other types of construction, the building of a home generally involves a variety of employers. Thus an understanding of the labor market must go beyond the homebuilder himself and encompass the market activities of the several

[10] For earlier studies in this regard, see Herbert R. Northrup, *Organized Labor and the Negro* (New York: Kraus Reprint Company, 1971), pp. 17-47; and Richard L. Rowan and Lester Rubin, *Opening the Skilled Construction Trades to Blacks*, Labor Relations and Public Policy Series, Report No. 7 (Philadelphia: Industrial Research Unit, The Wharton School, University of Pennsylvania, 1972).

[11] *Buffalo Evening News Almanac and Fact Book, 1973*, p. 35.

firms he engages to complete the work. In Erie County, a majority of homebuilders do not employ on-site labor except for supervisors and perhaps a few general tradesmen in order to rectify minor imperfections after the structure is finished. Most of the actual work, therefore, is performed by employees of subcontractors in the several functional areas. The "labor market," therefore, constitutes both the interactions between builders and subcontractors and those between both builders and subcontractors and their respective workers.

The basic methodological tool for this inquiry was a structured interview administered to both homebuilders and subcontractors in the following specialties: carpentry, cement work, electrical work, excavating and land clearing, drywall, heating and air conditioning, masonry, painting, plumbing, and roofing. Most of the interviews were conducted by the Survey Research Center at the State University of New York at Buffalo. The objective was to survey all builders and subcontractors operating in Erie County, although for reasons to be explained below the ultimate number of completed interviews fell somewhat short of this ideal.

The principal source of respondents for the survey was the Niagara Frontier Builders Association, the local chapter of the National Association of Home Builders (NAHB). The membership of this organization includes virtually all homebuilders in Erie County, but only a relatively small portion of subcontractors in the area. It was necessary, therefore, to solicit lists of subcontractors from the builders themselves. Although most of the builders interviewed were willing to provide the names and addresses of the subcontractors they used, some refused to do so and others specified firms or individuals who could not be traced. Through this procedure, a total of 251 employers were initially identified and contacted by letter; the breakdown by type is shown in Table 1.

For a variety of reasons, only 35.5 percent of the total builders contacted resulted in completed interviews. Over one-half of the homebuilders reported that they subcontracted all their work and, therefore, did not engage in any labor market activities. Thirty-three employers explicitly refused to be interviewed, and an additional thirty-four potential respondents either could not be reached by interviewers after repeated telephone calls or failed to appear for scheduled meetings. Fourteen more were primarily nonresidential contractors operating on a union basis,

TABLE 1. *Homebuilding Industry Employers Contacted, by Type Erie County, New York, 1971*

Type of Employer	Number
Homebuilders	104
Total homebuilders	104
Subcontractors:	
Carpentry	20
Cement	20
Drywall	5
Electrical	15
Excavating and land clearing	19
Heating and air conditioning	15
Masonry	11
Painting	15
Plumbing	21
Roofing	6
Total subcontractors	147
Total	251

Source: Erie County survey, 1971.

eight were out of business, eighteen worked by themselves without any employees, and one masonry contractor was insufficiently conversant in the English language to be interviewed. Finally, ten of those contacted (not included in Table 1) did not engage in on-site residential construction but in activities such as materials supply, prefabricated homes, remodeling, and real estate. The disposition of the employers and individuals shown in Table 1 is broken down further in Table 2.

The analysis of labor markets in residential homebuilding presented in the bulk of this report is based primarily on the 89 completed interviews indicated in Table 2. Since the survey was limited to Erie County, there was the danger that the results would be affected by characteristics of the area which were not typical of other communities. Although resources did not permit an extended canvassing of labor markets in different parts of the country, an attempt was made to solicit information in several other northeastern cities. In this regard, in-

TABLE 2. *Homebuilding Industry*
Disposition of Employers Contacted, by Type
Erie County, New York, 1971

Type of Employer	Total	Total Interviews Completed	Total Interviews Not Completed	Interviews Not Completed: All Work Subcontracted	Refused	Unable to Arrange Meeting	Nonresidential	Out-of-Business	Work Alone	Language Problem
Homebuilders:										
Total homebuilders	104	24	80	54	9	9	2	6	—	—
Subcontractors:										
Carpentry [a]	20	14	6		3	1	1	—	1	—
Cement [a]	20	11	9		3	6	—	—	—	—
Drywall	5	3	2		1	—	—	—	1	—
Electrical	15	5	10		4	3	—	1	2	—
Excavating and land clearing	19	7	12		4	4	2	—	2	—
Heating and air conditioning	15	9	6		2	3	—	—	1	—
Masonry	11	1	10		1	1	2	1	4	1
Painting	15	6	9		2	2	1	—	4	—

Plumbing	21	7	14		4	4	4	—	2	—
Roofing	6	2	4		—	1	2	—	1	—
Total subcontractors	147	65	82		24	25	12	2	18	1
Total	251	89	162	54	33	34	14	8	18	1

Source: Erie County survey, 1971.

[a] Includes some firms which also operate as builders.

terviews were conducted by the author in Rochester, Syracuse, Albany, Boston, Providence, and Cleveland. Those interviewed invariably included a spokesman from each of the local builders associations, as well as a few knowledgeable builders suggested by the association officials. Additional interviews were also conducted for the author in 1973 in the Philadelphia, Baltimore, Indianapolis, and Denver areas by the staff of the Industrial Research Unit. These interviews varied in length and intensity, but included association personnel and builders. The purpose was to test the generality of the conclusions reached as a result of the Erie County survey. The general study of open shop construction referred to in the Foreword will further determine the generality of the results reached in this study.

Published data on various aspects of homebuilding labor markets are exceedingly sparse, so that detailed comparisons of Erie County with areas other than those visited (or with the industry in the aggregate) are frequently impossible. There is some superficial information, however, in the periodic membership surveys made by the National Association of Home Builders which will be used for comparative purposes wherever appropriate.[12] Additional material on a nationwide basis was secured through interviews with various NAHB officials.

The discussions of residential labor markets in Chapters III-VI will include brief summaries and analyses of the available data on the union sector as represented in the existing literature. Some of the literature included contains material generated by previous research in connection with the author's doctoral dissertation.[13] This research includes interviews with over 70 business agents, contractors, association executives, welfare fund administrators, and government officials. Most of the investigation was concentrated in four upstate New York cities: Binghamton, Buffalo, Rochester, and Syracuse.

---

[12] The most recent NAHB membership survey referred to in this study is reported in Michael Sumichrast and Sara A. Frankel, *Profile of the Builder and His Industry* (Washington, D.C.: National Association of Home Builders, 1970).

[13] Foster, *op. cit.*; also, "Nonapprentice Sources of Training in Construction," *Monthly Labor Review*, Vol. XCIII, no. 2 (February 1970), pp. 21-26; "Labor Force Adjustments to Seasonal Fluctuations in Construction," *Industrial and Labor Relations Review*, Vol. XXIII, no. 4 (July 1970), pp. 528-540; "Apprenticeship Training in the Building Trades: A Sympathetic Assessment," *Labor Law Journal*, Vol. XXII, no. 1 (January 1971), pp. 3-12.

In addition, a mailed questionnaire was administered to union members of four crafts in the Syracuse area (bricklayers, carpenters, electricians, and operating engineers). A total of about 800 responses were received, constituting slightly less than 50 percent of the sample. The information gathered by the interviews and questionnaires was supplemented by examination of records from various welfare funds and apprenticeship programs. Most of the research for this study was conducted in 1968. The issues explored parallel much of the material covered in the homebuilding study, thus facilitating the comparative analysis to be presented in the ensuing chapters.

LIMITATIONS

It must be acknowledged at the outset that this report only begins to scratch the surface of the problem areas selected for investigation. No localized study with limited resources can hope to generate the breadth of information and analysis upon which firm and generalized conclusions can be based. More specifically, attention must be called to four factors which may tend to detract from the validity or utility of the results reported here. First, as with any survey, there is the problem of nonresponse. Although an attempt was made to identify all employers operating in Erie County, some were undoubtedly missed. Further, the large number of firms from whom interviews could not be obtained must be cause for concern. Although there is no explicit evidence that the experiences of those interviewed are in significant respects different from those who were not, this possibility does exist. Moreover, the total number of responses is too small to distinguish meaningfully the labor market experiences of the various types of employers surveyed.

In addition, the reader should be aware of the appreciable number of "contractors" who work alone and were thus excluded from the study. As will be seen, it is sometimes difficult to distinguish substantively between an employer-employee relationship and a contractor-subcontractor relationship. The survey did not solicit information on the process by which subcontractors are retained by homebuilders. Although it may well be that this process is in important ways similar to that by which employers and employees are matched, a material component of the residential construction labor market may have been omitted altogether.

A third limitation is the extent to which Erie County is not representative of markets in other regions of the country, or even of other areas in the Northeast. To the degree that particular factors are readily identifiable, either intuitively or through the already noted interviews with industry spokesmen in other cities and at the national level, they will be explicitly addressed. Nonetheless, in the absence of comprehensive nation-wide data, the findings of any local survey must be interpreted cautiously and allow for the possibility of differences in conditions elsewhere.

Finally, it must be emphasized that our information on residential construction was gathered exclusively from *employers.* Thus some important questions concerning the labor market experience of *workers* cannot be answered at all, and others can be approached only on the basis of second-hand responses. Had resources permitted us to supplement the employer interviews with input from employees, we might have learned a good deal more about hiring patterns, the impact of seasonality, and most particularly, skill acquisition.

OVERVIEW

Labor markets in construction are exceedingly complex. On perhaps the most basic level, the amount of movement in and out of the industry's work force, as well as within it, is staggering. The number of individuals who receive income from construction in a year is double the average monthly employment.[14] A comprehensive analysis of this movement would have to encompass a prodigious variety of interactions: between construction and other industries; among different local labor markets; between the skilled and the unskilled; among the various crafts; between the union and nonunion sectors; and among the residential, commercial, industrial, highway, and other industrial subdivisions. In addition, of course, there are the myriad of elements of labor market behavior—hiring, training, upgrading, wage determination, and so on—*within* each of these constituent parts. Even if we had available the statistical material necessary to include all these elements, which we do not, it could hardly be accomplished in a single volume.

The foregoing is intended merely to indicate that we begin this volume with an acute sense of diffidence. In no way do the

[14] Mills, *op. cit.*, p. 94.

following chapters offer a complete accounting of construction labor markets generally or those in homebuilding particularly. There are serious gaps, especially in the area of mobility patterns of construction workers.

> Unfortunately, the nature of these manpower flows is very imperfectly understood. We need longitudinal studies of individual workers which would help us answer such questions as: how do various categories of workers pick up their skills (through formal training or otherwise)? How do highly mobile workers differ from those who are less mobile? Especially, what are the work histories, job hunting techniques, etc., of men who spend only part of their work life in construction? [15]

The coverage here then is selective, although the issues raised are important ones. It should also be emphasized that with the small amount of information about labor markets in the union sector, there is even less information about their operation where unions play a minor role. This volume, therefore, should be seen as a modest foray into what is essentially virgin territory, one which will provide a starting point for subsequent efforts.

The following chapter will outline the fundamental economic characteristics of the industry and its labor force, as well as the extent and implications of the relative absence of unions in residential construction. Chapters III through VI cover in turn the four main areas explored in our field research: short-run manpower adjustments, long-run manpower adjustments, hiring procedures, and wages and their determination. The final chapter will summarize our findings, present our conclusions, and outline further research being conducted by the Industrial Research Unit which will supplement this volume.

---

[15] Howard G. Foster and George Strauss, "Labor Problems in Construction: A Review," *Industrial Relations*, Vol. XI, no. 3 (October 1972), p. 293.

CHAPTER II

Residential Construction

Many of the salient characteristics of the construction industry and its labor market are, by and large, reflections of the nature of the construction product. Although there is great variation in the kinds of facilities produced by the industry, all construction products exhibit a number of common properties from which devolve the structure and organization of the industry and, derivatively, the problems associated with labor force adjustment. The matter is put succinctly by Colean and Newcomb:

> The products of construction, despite the wide diversity already revealed, share certain common characteristics: they are immobile, complex, durable, costly, and speculative in value. The degree to which these characteristics are shared differs among the various kinds of structures and facilities; but in some degree they are universally present and affect all the other considerations with which we must deal.[16]

In one sense, there is no "market" for most kinds of construction products. With the exception of some single-unit residences, virtually all construction is "made to order." Work is started at the behest of the buyer, not the producer, and the specifications are tailor-made to the individual project. "In other words, the decisions as to what will be produced and when it will be produced come mainly from sources external to the industry." [17] The result is an extreme heterogeneity of demand, a condition necessitating a kind of industry organization which is capable of adjusting rapidly to sharp fluctuations in organizational and manpower requirements.

The need for flexibility, together with the vast heterogeneity and complexity of construction products, has led to a system of

---

[16] Miles L. Colean and Robinson Newcomb, *Stabilizing Construction: The Record and Potential* (New York: McGraw-Hill, 1952), p. 22.

[17] *Ibid.*, p. 81.

functional specialization. In the typical case, a buyer engages a general contractor (homebuilder) who has overall responsibility for the project. The general contractor, in turn, may engage any number of specialized contractors who perform certain specific operations such as wiring, plumbing, or painting.[18] Such an arrangement is indispensable in light of the nature of the product. To draw again from Colean and Newcomb's excellent analysis:

> Without the large and almost infinitely diverse aggregation of special trade contractors, the construction industry could not function under its existing requirements. This group provides the specialization within flexibility that permits the industry to accommodate itself to any assignment. The subcontractors, moreover, make possible the rapid mobilization and dismissal of job organizations necessitated by the changes both in the kind and volume of demand.[19]

The construction product is also distinguished by its immobility. Most structures obviously cannot be erected in one place and transported to another; therefore, the mass-production techniques found in other goods-producing industries are largely absent.[20] The resulting market structure is predominantly a local one. To be sure, there are nationwide construction firms whose gross receipts run into the hundreds of millions of dollars, but these typically undertake only the most extensive and costly projects, leaving the small- and medium-sized jobs, the bulk of all construction work, to local contractors. In regard to labor market problems, even the large national contractors must employ local labor and operate under locally determined conditions.

Finally, the durability and costliness of construction products, along with their particular vulnerability to national economic forces, render the demand for them highly unstable. The replacement or renovation of most existing structures can be deferred to a time when the economic horizons appear brightest. As a result, the demand for construction tends to fluctuate

---

[18] Of course, the general contractor may perform some of these functions with his own work force, although he will seldom if ever perform them all.

[19] Colean and Newcomb, *op. cit.*, p. 97. This is true even of small-scale residential construction.

[20] This characteristic has been tempered by the growing use of prefabricated materials which allow certain portions of the buildings to be produced on what approaches an "assembly-line" basis. In addition, large-scale housing tracts utilizing one basic plan may permit a kind of "factory" approach.

more dramatically than for other products, imposing on the industry a need for flexibility much greater than that required for other industries. Moreover, it creates an element of uncertainty which magnifies the problems otherwise inherent in developing and maintaining a skilled labor force.

Most firms in contract construction are small. According to the *1967 Census of Construction Industries,* the average firm with payroll employed slightly over nine persons.[21] The figure was even lower for the special trades contractor (nearly two-thirds of all firms) who averaged about seven workers. Special trades contractors account for about one-half of all contract construction employment, and even more among firms specializing in residential work.[22] Entry into the industry is relatively easy, and the incidence of business failure is high. Since World War II the number of construction firms has increased dramatically.[23]

THE STRUCTURE OF THE HOMEBUILDING INDUSTRY

Certain pertinent organizational characteristics of the homebuilding industry and its relationship to other sectors of construction will be outlined in this section. It must be noted at the outset that most available data series do not allow precise identification of homebuilding as a distinct entity within the construction industry. The Standard Industrial Classification delineates four major categories of construction firms: general building contractors, heavy and highway contractors, special trades contractors, and operative builders.[24] The homebuilding sector is represented in all of them except the heavy and highway contractors category and is usually not distinguished

---

[21] U.S. Bureau of the Census, *1967 Census of Construction Industries* (Washington, D.C.: Government Printing Office, 1970), Vol. I, p. 1A-2.

[22] *Ibid.*, pp. 26-3ff.

[23] For evidence on this point, and a detailed discussion of construction economics see Peter J. Cassimatis, *Economics of the Construction Industry,* Studies in Business Economics, No. III (New York: The Conference Board, 1969).

[24] Operative builders are firms, mostly residential, which produce structures for later sale—i.e., not on contract. Although this classification may for some purposes serve as a rough approximation of the "homebuilding industry," it represents only a fraction of it, the rest falling into the "general building contractors" category along with nonresidential builders.

statistically from nonresidential construction. Thus most statistical series published by public agencies do not offer insights into the character of the homebuilding industry.

Our purpose here is to construct a mosaic drawn primarily from three sources, which will offer a general picture of the economics of homebuilding and its influence upon labor market operations. The three sources are the interview survey administered in Erie County; the National Association of Home Builders membership survey [25] (hereafter, "NAHB"); and the *1967 Census of Construction Industries* [26] (hereafter, "Census"), the first comprehensive census of the construction industry since 1939.

Size and Nature of Firms

As with construction generally, the residential homebuilding sector is characterized by a large number of firms, most of them small in terms of both output and employment. The trend, however, appears to be in the direction of increasing concentration.[27] For example, in the most recent NAHB membership survey (1969), the median output for single-family builders was 21.8 units, up from 18.0 units in 1964 and 15.5 units in 1959.[28] Builders producing over 100 units rose from 6.2 percent of the total in 1959 to 12.7 in 1969. At the other end of the distribution, builders of 25 units or less declined from 69.5 percent to 57.5 percent. The complete distribution for the three survey years is shown in Table 3. (Among multifamily builders, the median in 1969 was 113.7 units, and for builders of both single and multifamily residences it was 25.5 units.)[29]

Despite the large number of firms, the industry does exhibit signs of concentration. The relatively few large firms account

---

[25] This membership survey is found in Michael Sumichrast and Sara A. Frankel, *Profile of the Builder and His Industry* (Washington, D.C.: National Association of Home Builders, 1970).

[26] U. S. Bureau of the Census, *op. cit.*

[27] John P. Herzog, "Structural Changes in the Homebuilding Industry" in William L.C. Wheaton *et al.* (eds.), *Urban Housing* (New York: Free Press, 1966).

[28] Sumichrast and Frankel, *op. cit.*, p. 108.

[29] *Ibid.*, p. 129.

TABLE 3. *Homebuilding Industry*
Percent Distribution of Single-Family Builders and Units
By Number of Units Built
United States, 1959, 1964, and 1969

Size of Builder by Number of Units Built	1959		1964		1969	
	Builders	Units	Builders	Units	Builders	Units
1-5	22.2	2.0	18.5	1.1	13.5	0.5
6-10	19.4	4.6	18.9	3.5	17.1	2.2
11-25	27.9	14.9	27.0	11.2	26.9	7.5
26-50	15.6	17.5	16.9	14.8	17.6	10.6
51-75	5.5	10.2	6.5	9.4	6.6	6.9
76-100	3.2	8.3	4.2	8.5	5.6	8.2
101-250	4.6	23.6	5.5	22.2	8.1	21.8
251-500	1.1	10.9	1.7	14.7	3.1	18.4
Over 500	0.5	8.1	0.8	14.7	1.5	23.9
Total	100.0	100.0	100.0	100.0	100.0	100.0
Median	15.5		18.0		21.8	

Source: Michael Sumichrast and Sara A. Frankel, *Profile of the Builder and His Industry* (Washington, D.C.: National Association of Home Builders, 1970), p. 108.

Note: Percents may not add to 100.0 due to rounding.

for a widely disproportionate percentage of the total units built.[30] (See Table 3.) In 1969, for example, one-eighth of the single-family firms produced over five-eighths of the homes. Again, this concentration has increased over time. In 1959, the largest, 14.9 percent of the firms, accounted for 61.1 percent of the total units.

The Erie County survey shows a roughly similar picture. Of 78 builder-respondents, 47 or 60.3 percent reported that they had built 25 units or fewer in the twelve months prior to the interview.[31] Only one builder in the county produced over 500 units. This distribution is shown in Table 4.

It is difficult to make precise statements as to employment levels among homebuilding firms, since each of our three sources presents data in somewhat different forms. Nonetheless, it is sufficient for present purposes to highlight the fact that these firms generally are not large employers. The 1967 Census, for example, shows that general building contractors specializing in single-family houses averaged just under four employees, with multifamily builders averaging a little over ten workers.[32] The Census also reveals that operative builders averaged about five and one-half employees, two-thirds of whom were on-site construction workers.[33] These figures are reached by averaging employment levels in March, May, August, and November. The NAHB survey, on the other hand, refers only to "full-time employees on payroll year round," but nevertheless reports a somewhat higher figure. Both the Census and NAHB data are limited to firms with at least one employee. The NAHB figures are presented in Table 5.

The NAHB report does not contain information on the distribution of employment, although it does reveal that employers producing over 100 units averaged roughly four times as many

[30] At the same time, large builders are not necessarily insulated from the vagaries of the market by their size, and some of them even failed to survive the industry's slump in late 1973 and early 1974. See "An Alarming Fall for Two Top Builders," *Business Week*, January 5, 1974, pp. 22-23; and "U.S. Home's Big Mistake," *Business Week*, January 12, 1974, pp. 44-45.

[31] It should be noted that the 78 builders include those who put up some multiunit structures. As we shall see, the number performing multifamily work exclusively is very small, and even they are not the largest builders in the group.

[32] U.S. Bureau of the Census, *op. cit.*, Vol. I, p. 26-3.

[33] *Ibid.*, p. 24-6.

TABLE 4. *Homebuilding Industry*
Size of Homebuilders, by Annual Number of Units Built
Erie County, New York, 1971

Size of Builder by Number of Units	Number of Homebuilders	Percent
0	3	3.8
1-10	22	28.2
11-25	25	32.1
26-50	12	15.4
51-100	8	10.2
101-250	7	9.0
251-500	—	—
Over 500	1	1.3
Total	78	100.0

Source: Erie County survey, 1971.

TABLE 5. *Homebuilding Industry*
Full-time Employment, by Type of Employee
United States, 1969

Type of Employee	Number of Employees: Mean	Number of Employees: Median
Executive and professional	2.8	1.5
Office personnel	3.4	1.4
Salesmen	3.4	1.6
Construction supervisors	2.2	1.3
Carpenters	7.6	4.5
Masons	5.0	1.9
Painters	3.2	1.7
Laborers	6.9	1.8

Source: Michael Sumichrast and Sara A. Frankel, *Profile of the Builder and His Industry* (Washington, D.C.: National Association of Home Builders, 1970), Table 57, p. 201.

workers as those building 25 units or less.[34] However, the Census does provide such data for operative builders (see Table 6) which show again the concentration of employment among the relatively large employers. Although 4.6 percent of the firms accounted for 39.0 percent of the employment, the most striking observation is that, in all the United States, only 49 operative builders employed 100 or more people.

Again, the picture is similar in Erie County. The builders were asked for the number of their employees at the time of the interview (primarily during the fall of 1971) and the number they would have on their payroll during the "busy" season. The questions referred specifically to on-site workers, therefore, office, technical, and sales personnel are excluded. Table 7 summarizes the responses to these two questions.

Despite the welter of statistics from different sources with different exclusions and definitions, the conclusion that employment among homebuilders is thoroughly dispersed is inescapable. As noted earlier, however, the homebuilders themselves do not constitute the "industry"; indeed, the lion's share of on-site employment is generated through subcontractors. Unfortunately, the NAHB report does not offer any information on employment patterns among subcontractors. Certain clues can be gleaned from census data with somewhat more detail provided by the Erie County survey. The Census permits calculation of averaged employment levels for various kinds of subcontractors specializing[35] in residential construction (Table 8), however, it should be emphasized that these figures do not take account of subcontracting firms or individuals without payrolls.

The averages in Table 8 demonstrate again the highly dispersed character of residential construction employment, particularly among subcontractors specializing in single-family homes. Obviously, the figures would be even lower if those firms without any employees were included. Furthermore, it is worth noting that average employment tends to rise as the degree of specialization in residences declines. For example, plumbing and heating contractors who worked exclusively in single-family homes averaged 2.8 workers, but those with 51

---

[34] Sumichrast and Frankel, *op. cit.*, p. 201.

[35] The term "specializing" used here and elsewhere in this chapter refers to 51 percent or more of the firm's receipts in the area of specialization.

TABLE 6. ***Construction Industry***

Number and Percent of Establishments and Employees, by Size of Establishment
United States, 1967

Size of Establishment by Number of Employees	Establishments		Employees	
	Number	Percent	Number	Percent
1-4	9,467	71.5	16,438	22.7
5-9	2,165	16.4	13,347	18.5
10-19	1,005	7.6	14,301	19.8
20-49	476	3.6	13,744	19.0
50-99	75	0.6	4,817	6.7
100-248	39	0.3	5,805	8.0
250 or more	10	0.1	3,853	5.3
Total	13,237	100.0	72,305	100.0

Source: U.S. Bureau of the Census, *1967 Census of Construction Industries* (Washington, D.C.: Government Printing Office, 1970), Vol. I, p. 24-5.

Note: Percents may not add to 100.0 due to rounding.

TABLE 7. *Homebuilding Industry On-Site Employment, by Size of Builder Erie County, New York, 1971*[a]

Size of Builder by Number of Employees	Number of Builders			
	At Time of Interview		During Busy Season	
	Number	Percent	Number	Percent
0	1	3.1	—	—
1-4	13	40.6	10	31.2
5-9	12	37.5	11	34.4
10-19	2	6.3	7	21.9
20-49	4	12.5	3	9.4
50-99	—	—	1	3.1
100 or more	—	—	—	—
Total[b]	32	100.0	32	100.0

Source: Erie County survey, 1971.

[a] Includes only those builders who perform on-site work on their own.

[b] Includes 23 builders (one builder did not answer) plus 9 carpentry and cement contractors who also acted as builders.

to 59 percent specialization averaged 7.5 employees.[36] On the other hand, firms specializing in multifamily residences tend to exhibit employment levels roughly similar to nonresidential construction.[37]

The Erie County survey permits us to go beyond simple averages and look at distributions. Again the reader is reminded that these data are limited to on-site construction workers, and that they refer not to an annual average but to employment at the time of the interview or during the "busy" season. Nevertheless, it is sufficient to note for present purposes that the picture of a dispersed, highly atomistic industry is corroborated (see Table 9). Approximately two-thirds of the respondents had fewer than 10 workers and only about one-sixth had 20 or more. Although the average level suggested in Table 9 is somewhat higher than those revealed in the Census,

[36] U.S. Bureau of the Census, *op. cit.*, p. 26-8.

[37] *Ibid.*

TABLE 8. *Construction Industry*
Average Employment of Subcontractors
Specializing in Residential Construction, by Type
United States, 1967

Type of Subcontractor	SIC	Average Number of Workers: Single-Family	Average Number of Workers: Multifamily
Plumbing, heating, air conditioning	1711	4.0	10.9
Painting, paper hanging	1721	2.8	5.6
Electrical	1731	3.4	10.2
Masonry	1741	3.9	15.8
Plastering	1742	3.6	14.5
Terrazzo, etc.	1743	3.5	5.9
Carpentry	1751	3.7	12.9
Floor laying	1752	3.7	6.6
Roofing and sheet metal	1761	4.4	5.3
Concrete work	1771	4.7	13.1
Glass work	1793	3.0	11.8
Excavating	1794	4.1	6.8

Source: Derived from: U.S. Bureau of the Census, *1967 Census of Construction Industries* (Washington, D.C.: Government Printing Office, 1970), Vol. I, pp. 26-8ff.

TABLE 9. *Homebuilding Industry*
On-Site Employment, by Size of Subcontractor
Erie County, New York, 1971

Size of Subcontractor by Number of Employees	Subcontractors At Time of Interview: Number	At Time of Interview: Percent	During Busy Season: Number	During Busy Season: Percent
0	2	3.1	—	—
1-4	17	26.6	16	25.0
5-9	23	35.9	21	32.8
10-19	12	18.8	16	25.0
20-49	10	15.6	10	15.6
Over 50	—	—	1	1.6
Total [a]	64	100.0	64	100.0

Source: Erie County survey, 1971.

[a] Total includes all interviewed subcontractors minus one no response.

it is worth noting that the Erie County survey did not uncover all employers in the county, and it is likely that those missed are relatively small. Thus we may conclude that the market structure in this area does not seriously diverge from that exhibited on a nationwide basis.

Other Structural Characteristics

An important question with respect to labor markets in homebuilding is the degree to which these markets dovetail with those in other sectors of the construction industry. The question is important for a number of reasons. The extent to which the residential sector influences and is in turn influenced by conditions in the more heavily unionized commercial and industrial sectors must be considered along with the competitive threat posed by a large nonunion work force to the standards established by the building trades unions. The relationship among the sectors is a complicated one, and we shall be referring to it throughout the following chapters. Here, however, we shall limit ourselves to those aspects which permit relatively "hard" conclusions.

The preponderance of builders who specialize in housing operates primarily in the single-family market. The NAHB survey reveals that nearly 40 percent of the respondents had their "primary operation" in single-family homes, with less than 10 percent in multifamily residences. Of those identified as *both* single and multifamily firms, nearly two-thirds were chiefly single-family.[38] Furthermore, it should be noted that there may have been a bias in the survey, since the smaller builder, generally single-family, is less likely to have responded or even to have been a member of the NAHB.

This conclusion is supported by census data which show over 57,000 general building contractors specializing in single-family homes and less than 3,000 in multifamily.[39] (See Table 11.) Likewise, it was found that in Erie County 70.5 percent of the builders produced more than three-fourths of the units as single-family homes (Table 10). In short, the bulk of the residential construction industry is comprised of firms which produce predominantly single-family housing units.

---

[38] Sumichrast and Frankel, *op. cit.*, p. 91.

[39] U.S. Bureau of the Census, *op. cit.,* Vol. I, p. 26-3.

TABLE 10. *Homebuilding Industry*
Homebuilders by Percent of Units That Are Single-Family
Erie County, New York, 1971

Percent of Units That Are Single-Family	Number of Builders	Percent
0-25	12	15.4
26-50	7	9.0
51-75	4	5.1
76-99	24	30.8
100	31	39.7
Total	78	100.0

Source: Erie County survey, 1971.

Single-family builders are less likely to diversify into other fields than are other types of contractors. According to the 1967 Census, 70 percent of the single-family specializers did not diversify as compared to less than one-half of the multifamily specializers.[40] Both percentages are higher if operative builders are added to the general contractors group. The Census does not reveal whether the secondary operations of single-family builders are in multifamily housing or outside residential construction altogether. The NAHB survey does show, however, that 11.9 percent of the single-family builders did some work in commercial or industrial construction. The comparable percentages for multifamily builders and single-multi combined were 25.7 and 27.9, respectively.[41] In Erie County, 49 of the 73 builders or 67 percent responding to the question report *no* business outside residential construction, with another 15 builders or 21 percent reporting one to 5 percent of their business outside housing (see Table 12). In summary, *among builders,* there appears to be a fairly distinct housing sector, with most firms (especially single-family, the most numerous) concentrating more heavily in their sphere of specialization.

The picture is somewhat different if one inquires into the practices of subcontractors because the demarcation between resi-

[40] *Ibid.*

[41] Sumichrast and Frankel, *op. cit.*, p. 90.

TABLE 11. *Construction Industry*
Percent Distribution of Specialization Among Contractors and Subcontractors with Payroll Specializing in Single-Family and Multifamily Residences
United States, 1967

Type of Firm	Single-Family Dwellings							Multifamily Dwellings						
		Degree of Specialization							Degree of Specialization					
	Number	100	90-99	80-89	70-79	60-69	51-59	Number	100	90-99	80-89	70-79	60-69	51-59
Plumbing	20,804	31.0	25.4	15.5	15.4	10.7	2.0	1,606	22.8	16.9	16.4	17.7	21.1	5.2
Painting	14,819	48.1	21.4	12.3	10.7	6.0	1.4	1,647	51.2	11.3	13.0	11.3	11.5	1.7
Electrical	7,491	21.3	19.9	18.1	19.3	16.2	5.1	888	10.4	16.8	18.0	25.1	23.4	6.3
Masonry	11,303	57.3	20.3	8.7	8.6	4.3	0.7	740	35.0	12.4	11.8	19.9	18.6	2.3
Plastering	2,613	43.2	24.3	8.4	14.2	8.5	1.4	272	37.1	14.3	11.4	6.3	16.9	14.0
Terrazzo	2,692	35.8	27.3	16.7	13.4	6.4	0.4	237	10.5	16.5	14.8	35.0	17.7	5.5
Carpentry	13,025	69.5	12.2	9.4	5.0	3.4	0.7	1,027	42.6	18.2	10.2	13.0	8.2	7.9
Floor laying	4,083	23.3	30.8	18.8	17.8	7.2	2.1	268	17.5	6.7	16.8	17.2	40.3	1.5
Roofing and sheet metal	6,200	35.5	23.3	14.4	14.2	9.2	3.4	375	12.8	2.9	27.2	25.6	30.1	1.3
Concrete	5,320	49.3	19.1	12.7	11.2	5.9	1.6	524	47.7	17.0	5.0	13.4	11.8	5.2
Glass	269	24.2	23.4	8.9	20.4	22.7	0.4	35	2.9	28.6	—	2.9	62.9	2.9
Excavating	2,717	37.9	18.6	11.2	19.7	9.3	3.3	141	17.7	19.1	13.5	22.7	17.7	9.2
General builders	57,343	70.1	10.7	6.9	6.1	4.8	1.4	2,943	49.3	11.0	11.2	12.2	13.0	3.1
Operative builders	11,852	90.1	5.4	1.4	1.5	1.3	0.3	630	72.1	4.4	7.1	7.8	7.6	1.0

Source: Derived from: U.S. Bureau of the Census, *1967 Census of Construction Industries* (Washington, D.C.: Government Printing Office, 1970), Vol. I, pp. 26-3ff.

Note: Percents may not add to 100.0 due to rounding.

TABLE 12. *Homebuilding Industry*
Number and Percent of the Proportion of Single-Family and Nonresidential Construction Work, by Builders and Subcontractors Erie County, New York, 1971

Percent	Builders		Subcontractors	
	Number	Percent	Number	Percent
Single-family:				
0-25	12	15.4	7	10.8
26-50	7	9.0	14	21.5
51-75	4	5.1	16	24.6
76-99	24	30.8	24	36.9
100	31	39.7	4	6.2
Total	78	100.0	65	100.0
Nonresidential:				
0	49	67.1	25	39.1
1-5	15	20.5	17	26.6
6-10	5	6.9	12	18.7
11-20	3	4.1	7	10.9
21-50	1	1.4	3	4.7
50-100	—	—	—	—
Total [a]	73	100.0	64	100.0

Source: Erie County survey, 1971.

[a] Total includes all builders minus five no response and all subcontractors minus one no response.

dential and nonresidential is less clear. As Table 11 shows, subcontractors are considerably more likely to perform work outside their area of specialization than are builders. Certain other patterns emerge from the table as well. In virtually every case, multifamily firms specialize less than single-family firms, suggesting again that there is more overlap between the multifamily residential sector and nonresidential construction than between single-family and nonresidential. The reason undoubtedly is that the managerial and capital resources necessary to initiate large projects inhibit the single-family specializer from moving outside his specialty, whereas the multifamily firm is more likely to have those resources necessary to move into nonresidential construction as well as single-family building. Furthermore, specialization appears to be greatest among subcontractors who perform the basic structural work (carpentry, masonry, and concrete) and least among the mechanical (electrical, plumbing) and finishing (floor laying, terrazzo) industries, although exceptions to these generalizations are evident (e.g., painting). Finally, there is a slight tendency for subcontractors to specialize in multifamily building more than the general contractors, with the numerical ratio (multi:single) roughly 1:10 for the former and 1:20 for the latter.

These generalizations are applicable to Erie County as well. Subcontractors were less prone than builders to specialize in single-family housing and more likely to perform work outside the residential sector, as illustrated in Table 12. Unfortunately, there are not enough observations for each type of subcontractor to permit detailed breakdowns similar to those in Table 11. Nonetheless, it appears safe to conclude that the nexus between the residential and nonresidential labor markets (and derivatively that between the union and nonunion sectors) is stronger for subcontractors than for builders. Thus it is appropriate to inquire at this point into the extent and nature of subcontracting and the relationship between the builder and his subcontractors.

SUBCONTRACTING IN RESIDENTIAL CONSTRUCTION

According to the 1967 Census, about one-third of the total receipts of general building contractors specializing in single-family homes was paid out to subcontractors and the proportion for multifamily firms was over one-half.[42] The corresponding figures for operative builders were 40 percent and 56 percent, respectively.[43] In all cases, the amount paid out to subcontractors exceeds the value added by the builder. The NAHB survey does not provide an overall ratio, but its tabular breakdown suggests an average which is somewhat higher than that of the Census.[44]

---

[42] U.S. Bureau of the Census, *loc. cit.*

[43] *Ibid.*, p. 26-32.

[44] Sumichrast and Frankel, *op. cit.*, p. 162. The discrepancy may be attributed to NAHB's use of "construction costs" as the denominator rather than "receipts." The term "costs" is not defined.

However, the NAHB also presents its findings from earlier surveys which indicate clearly that the proportion of work subcontracted has risen steadily since 1959 (see Table 13). Regardless of the precise proportion of costs or receipts allocated to subcontractors, the subcontractors provide a much larger percentage of total employment, and the NAHB figures suggest that this percentage has been rising.

The Erie County survey did not solicit information on the *amount* of work subcontracted, but it does provide insights into the kind of work subcontracted and the nature of the subcontracting relationship. The survey identified 12 building functions and asked the homebuilder whether he did each type of work himself, subcontracted, or did both. The total number of respondents on this question is 86 (as opposed to 78 previously), as it includes 8 firms which act as *both* builders and subcontractors. The responses are shown in Table 14. A response that work was done by both the builder and a subcontractor generally meant a distinction between particular kinds of work (e.g., rough vs. finished carpentry), although in some cases it meant simply that the builder would do it himself if his work load permitted.

The figures in Table 14 generally reflect the fact that some two-thirds of the builders subcontracted *all* their work. Interviews with industry spokesmen in other northeastern cities revealed that the practice is common and that in this respect Erie County is not unrepresentative of the industry as a whole in the Northeast. Even when some work is performed by the builder himself, certain jobs are almost invariably subcontracted (heating, plumbing, electrical, drywall). Except for carpentry, there appears to be no type of work which is still performed by a substantial number of builders.

The relationship between builders and subcontractors is not the same in all instances. The differences may be perceived as lying along a spectrum. At one extreme, the subcontractor is a well-established firm which provides services and often materials to a wide variety of contractors both within and without the residential sector. This type of subcontractor may also perform a substantial amount of repair and remodeling work either under a direct arrangement with the purchaser or through a contractor. At the other extreme, the subcontractor is an individual (or small group) who may work primarily for a single builder, sometimes one for whom he previously worked as an

TABLE 13. *Homebuilding Industry*
Percent Distribution of Cost Subcontracted
United States, 1959, 1964, and 1969

Percent of Cost Subcontracted	Number of Respondents		
	1959	1964	1969
0-24	19.4	13.8	11.9
25-49	22.8	22.2	19.7
50-74	27.1	26.5	28.3
75-100	30.7	37.5	40.0

Source: Michael Sumichrast and Sara A. Frankel, *Profile of the Builder and His Industry* (Washington, D.C.: National Association of Home Builders, 1970), p. 162.

TABLE 14. *Homebuilding Industry*
Type of Work Performed by Builder and Subcontractor
Erie County, New York, 1971

Type of Work	Total [a]		Builder		Subcontractor		Both	
	No.	Percent	No.	Percent	No.	Percent	No.	Percent
Carpentry	86	100.0	20	23.2	55	64.0	11	12.8
Cement	86	100.0	6	7.0	74	86.0	6	7.0
Drywall	86	100.0	2	2.3	82	95.4	2	2.3
Electrical	86	100.0	1	1.2	84	97.6	1	1.2
Excavation	86	100.0	2	2.3	82	95.4	2	2.3
Heating	86	100.0	2	2.3	83	96.5	1	1.2
Land clearing	86	100.0	12	14.0	71	82.5	3	3.5
Masonry	86	100.0	4	4.7	79	91.8	3	3.5
Painting	86	100.0	7	8.1	75	87.2	4	4.7
Plumbing	86	100.0	—	—	85	98.8	1	1.2
Roofing	86	100.0	4	4.7	77	89.5	5	5.8

Source: Erie County survey, 1971.

[a] Total includes 8 firms which act as both builders and subcontractors.

employee. These subcontracting "firms" are frequently ephemeral, employ few workers of their own, and bear a relationship with a builder which roughly approximates that of an employee working piece rate. Of course, there are many subcontractors who fall between the two extremes. There are large, well-established firms which do most or all of their work with a single builder, just as there are short-lived individual or small group businesses which perform services for many builders.

The nature of the relationship between builder and subcontractor is likely to depend on the kind of work involved. If the operation requires considerable expertise or licensing or requires the subcontractor to provide his own equipment and materials, the relationship is apt to be closer to the contractor-subcontractor pattern than to the employer-employee one. If the operation is relatively uncomplicated, requires little investment in equipment and material, is amenable to piecework, and is one which the builder *could* easily do himself (or has in the past done himself), then the subcontractor is more likely to be an individual or small group with a close working relationship with a single builder. This distinction is intended only to portray a general pattern and is by no means absolute.

The Erie County survey, as well as the interviews with other industry spokesmen, revealed two fairly discernible groups of functions. Operationally, the distinction is made on the basis of three questions on the survey: (1) What is the percentage of work other than new building? (2) Does the subcontractor complete over one-half his work with a single builder? and (3) Does the subcontractor stock and sell building supplies as well as labor? Group I includes carpentry, masonry, and painting, whereas Group II includes cement work, excavation and land clearing, electrical, plumbing, heating, roofing, and drywall. Table 15 summarizes the responses to the three above-mentioned questions.

Generally, Table 15 illustrates the greater diversification among Group II subcontractors. They are more likely to work with different builders, to perform repair and remodeling work, and to stock and sell building materials. This diversification is usually a reflection of a stable, ongoing business establishment, with a steady core of its own employees, whose relationship with builders is relatively independent.

Group I subcontractors, on the other hand, are more specialized and are more likely to be only labor brokers. In Erie County, the research indicates that many subcontractors are established

TABLE 15. *Homebuilding Industry Selected Operational Characteristics of Group I and II Subcontractors Erie County, New York, 1971*

Characteristics	Group I [a]	Group II [b]
Work other than new:		
None	14	22
Some	7	22
One-half of work with single builder:		
Yes	10	4
No	11	40
Stock and sell materials:		
Yes	1	9 [c]
No	20	33 [c]

Source: Erie County survey, 1971.

[a] Total respondents equals 21.

[b] Total respondents equals 44.

[c] Two respondents did not answer.

in businesses by builders who once employed them and wish to free themselves of an employer's burdens, whereas others break away from the subcontractors who once employed them and form a competing business. These subcontractors frequently work alone, or with one or two helpers, and rely on word-of-mouth to inform builders of their availability. Furthermore, they invariably work out of their homes and often find it unnecessary even to maintain an office there. It is noteworthy that they are rarely to be found in the telephone book as business firms.

The importance of these observations is in regard to one of the limitations noted in Chapter I. For Group II subcontractors, the critical labor market questions involve primarily the interactions between the subcontractor and his employees. He is the employer in every sense of the word. For Group I subcontractors, however, the interactions between the builder and subcontractor assume greater significance; the subcontractor may be in certain respects an "employee" of the builder in all but form (particularly if the subcontractor himself has no workers). Thus, to the extent that Group I subcontractors are underrepresented in our survey because they were by nature difficult to locate (or excluded precisely because they had no workers), some important labor market questions cannot be answered as thoroughly as would be desirable.

THE LABOR FORCE

The construction labor force is difficult to define because of the industry's inordinate turnover; the numbers change dramatically depending on the particular measure used. In 1971, for example, average annual employment was about 3.4 million, but a total of 5.4 million individuals had their longest job in construction. This ratio of 1.6 compares to 1.2 for manufacturing and 1.3 for the economy as a whole.[45] The number of workers with any earnings in construction is in excess of 6 million.

As noted earlier, construction employment has kept pace with the growth of the work force, generally the only goods-producing industry to do so. Table 16 compares average annual employment in 1972 with the levels in 1948 for several industries.

The construction work force as a whole is comprised chiefly of blue collar workers, although the proportion has been slowly declining. One-half of the workers are classified as craftsmen and foremen. Although this percentage has remained fairly steady in recent years, there have been marked shifts in the specific craft composition of the skilled group as a result of changes in technology and in the composition of the construction product. Thus the relative employment levels of such trades as carpenter, painter, and plasterer have fallen, whereas those of the plumber, electrician, and operating engineer have increased. Table 17 offers a breakdown of the industry's labor force by major occupational group.

Although unemployment figures by industry are somewhat suspect,[46] there is little doubt that the construction worker suffers an inordinate amount of joblessness. According to the published data, the unemployment rate for construction has consistently

---

[45] U.S. Department of Labor, *op. cit.* pp. 184, 189.

[46] David L. Kaplan, "Unemployment by Industry—Some Comments on Its Measurement and Behavior" in National Bureau of Economic Research, *The Measurement and Behavior of Unemployment* (Princeton: Princeton University Press, 1967), pp. 281-289.

TABLE 16. *Selected Industries Average Annual Employment United States, 1948 and 1972*
(Numbers in thousands)

Industry	1948	1972	Percent Change
Agriculture	7,628	3,472	–54.5
Nonagriculture	44,891	72,750	62.1
Mining	994	607	–38.9
Contract construction	2,169	3,520	62.3
Manufacturing	15,582	18,928	21.5

Source: U.S. Department of Labor, *Manpower Report of the President* (Washington, D.C.: Government Printing Office, 1973), Tables A-13, C-1.

TABLE 17. *Construction Industry Percent Distribution of Employment By Major Occupational Group United States, 1959 and 1970*

Occupational Group	1959	1970
Professionals and technicians	4.4	4.6
Managers	11.9	11.8
Clerical and sales workers	4.9	5.7
Total white collar	21.2	22.1
Craftsmen	50.3	50.6
Operatives	8.7	9.9
Laborers	19.4	16.9
Service workers	0.4	0.4
Total blue collar	78.8	77.8
Total	100.0	100.0

Source: U.S. Bureau of Labor Statistics, *Labor Force and Employment in 1959*, Special Labor Force Report No. 4 (Washington, D.C.: Government Printing Office, 1959), Table C-9, and *Handbook of Labor Statistics 1971* (Washington, D.C.: Government Printing Office, 1971), Table 20.

Note: Percents may not add to 100.0 due to rebounding.

been double the all-industry average.[47] Construction is especially prone to most of the conventional classifications of unemployment. The industry's sensitivity to general economic conditions, both nationally and locally, results in a disproportionately high level of deficient-demand unemployment. Seasonal unemployment is among the highest of all industries, and employment turnover leads to substantial frictional unemployment.

Unfortunately, detailed employment figures are not available for homebuilding or residential construction alone. Although 40 percent of all construction *activity* is residential, data on the value of construction cannot be readily translated into labor force terms.[48] Nevertheless, certain clues can be obtained indirectly. According to the *1967 Census of Construction Industries*, some 286,000 workers were employed by firms classified as general building contractors or operative builders which specialized in the construction of single-family homes, or about 30 percent of the workers employed by all firms in these categories. Similarly, 369,000 workers were employed by special trades contractors specializing in single-family homes, or about 25 percent of total employment in that group.[49] These figures, however, can be regarded only as approximations of residential employment, since specializing firms do not necessarily perform all their work in single-family house construction, just as nonspecializing firms (or firms specializing in other work) may do *some* single-family building. Furthermore, we have not included data on multifamily dwellings or other forms of residential construction (dormitories, hotels, etc.), since we are primarily concerned here with the single-family market.

A crude occupational breakdown of single-family employment may be obtained in the following manner. In 1969, the Bureau of Labor Statistics surveyed the labor requirements of 250 single-family homes throughout the United States.[50] It was estimated that each $1,000 expenditure generates 62 man-hours of

---

[47] U.S. Department of Labor, *op. cit.*, p. 151.

[48] U.S. Department of Commerce, "Noncomparability of Statistics on Contract Construction Employment and Value of Construction Put-in-Place," *Construction Review*, Vol. I, no. 1, (March 1955), pp. 7-9.

[49] Derived from U.S. Bureau of the Census, *loc. cit.*

[50] Robert Ball and Larry Ludwig, "Labor Requirements for Construction of Single-Family Houses," *Monthly Labor Review*, Vol. XCIV, no. 9 (September 1971), pp. 12-14.

construction labor, consisting of 52 on-site and 10 off-site man-hours. The on-site labor requirements were broken down according to occupation and employing contractor and are reproduced in Table 18.

To obtain numerical estimates of employment by occupation, we begin with the total value of single-family construction in 1969, $16.4 billion, divide by 1,000 and multiply by 52. The total provides the number of on-site man-hours required. Dividing again by 2,000 will supply the total number of man-years, 426,-400.[51] It must be emphasized, however, that both approaches give only average employment, not the total number of workers in the industry. The turnover rates for homebuilding alone have not been provided.

Nature of Employment Relationships

The characteristics of construction operations lead to a situation where the attachment of workers to individual employers is relatively weak. Different kinds of projects require different work groups, and the work force on any given project changes as the structure enters succeeding stages of construction. Unless an employer is large and/or diversified, he may not be able to retain many workers whose phase of the project has been completed. As work groups are dissolved, workers must often find employment with a different contractor. It is not unusual for some individuals to work for several different employers in the course of a year.

This pattern is exacerbated by the seasonal volatility of construction work. Even large employers find it necessary to lay off workers in the winter months; consequently, these workers may not be available for reemployment in the spring. As a result, much of the market operates through a pool of workers from which employers draw according to their needs. In union construction, the "pool" may be administered through a hiring hall or referral service; in the nonunion sector, employers rely largely on word-of-mouth to inform workers in the pool that jobs are open.

---

[51] The Census figures include *off-site* employees of construction firms as well as some firms engaged in addition and alteration work, whereas the other data are limited to *new* housing units. Hence, we would except the Census employment levels to be somewhat higher, as they appear to be. We are concerned here, however, merely with the order of magnitude.

TABLE 18. *Homebuilding Industry Percent Distribution of On-Site Man-Hours, By Occupation and Type of Contractor United States, 1969*

Occupation	Percent Distribution of Man-Hours	Contractor	Percent Distribution of Man-Hours
Supervisory, professional, technical, and clerical	2.8	General	31.3
Carpenters	34.9	Carpentry	16.9
Painters	7.3	Plumbing, heating, ventilating, and air conditioning	8.7
Bricklayers	5.7	Painting and paper hanging	6.5
Plumbers	4.3	Masonry and stone work	9.0
Cement finishers	2.5	Concrete and stucco work	7.2
Electricians	3.0	Electrical (except heating)	3.7
Plasterers and lathers	1.7	Plastering and lathing	1.4
Sheet metal workers	1.3	Roofing and siding	2.0
Roofers	0.9	Ceramic tile, terrazzo, and marble work	2.2
Operating engineers	1.8	Excavation and grading	11.6
Tile setters	1.4	Wood flooring	0.9
Soft floor layers	0.6	Other flooring	1.6
All other skilled trades	3.3	Wallboard	4.0
Laborers	14.1	All other types	2.9
Helpers and tenders	13.8		
Truck drivers and miscellaneous workers	0.5		
Total	100.0	Total	100.0

Source: *Monthly Labor Review,* Vol. XCIV, no. 9 (September 1971), p. 13.

Note: Percents may not add to 100.0 due to rounding.

The foregoing generalizations do not apply equally to all segments of the industry. Some special trades contractors, and some small general contractors as well, maintain more or less permanent work forces, and in these cases the employment relationship is similar to that in other, more stable industries. Most contractors, moreover, employ a cadre of key workers who enjoy relatively steady employment. As we shall see in Chapter IV, although employment patterns and levels in homebuilding are indeed unstable, there is a sense in which this sector exhibits a relative steadiness that is uncommon to the industry as a whole.

In summary, it has been observed that the construction labor force is comprised of three distinguishable groups. The first group includes those workers who enjoy year-round employment, usually with a single employer. The second contains those workers attached to the industry but who follow the pattern of moving from employer to employer, often interrupted by spells of joblessness. The third is composed of workers with only a secondary attachment to construction who enter the industry for brief periods, usually during the busy season, and who return to their primary activity (often as students) when work ceases or school reopens. Any labor market analysis must take account of the different characteristics of these three groups.

INSTITUTIONAL ORGANIZATION

The fragmentation of the construction industry is reflected in a complex network of trade associations into which many employers are organized. These associations generally follow the various industrial divisions: there are several groups of general contractors which are sometimes divided between building construction and heavy and highway construction; and most of the special trade categories have their own associations, for example, the National Electrical Contractors Association, Painting and Decorating Contractors Association, Mechanical Contractors Association, etc. In the nonunion sector, groups such as the Associated Builders and Contractors and the National Association of Home Builders, which also include some union builders, contain both general and specialty contractors. The total number of associations in the industry is literally in the hundreds, and an individual employer may belong to several at one time.

The activities of national trade associations and their local affiliates, particularly those with unionized members, are summarized succinctly by Mills:

> The local associations perform a wide variety of functions for their members, including public relations, lobbying, legal advice, labor relations activities, and members' benefits (such as group life insurance for

contractors or types of liability insurance), and they deal with architects, owners, suppliers, and others. The national office of the association also conducts lobbying and public relations and provides legal and industrial relations advice. It often publishes periodicals carrying trade news, innovations, legislative reports, and analyses of the national scene as it affects members' concerns.[52]

In the union sector, the labor relations role of local associations is substantial, occupying much of the time of association executives and including representation of members in collective bargaining and contract administration.

The scope of an association's activities is usually related to the size of its staff. The National Association of Home Builders, for example, has a large national staff, thus performing functions in addition to those enumerated by Mills. The NAHB carries on an extensive research operation, and as we shall see, it administers manpower training programs under contract with the federal government. The Association estimates that in 1968 about one-half of all professionally built single-family units were constructed by its builder-members.[53]

The Balkanization of the industry extends to its union organization as well. Organized construction workers are represented by no fewer than nineteen national unions, and the industry's bargaining structure is incredibly complicated. For the most part, each local union negotiates separately with one or more local contractors' associations and often with individual nonassociation employers as well. In general, the "basic" trades (carpenter, laborer, operating engineer, etc.) negotiate with groups of general contractors, whereas the "specialty" trades (electrician, plumber, painter, etc.) negotiate individually with associations of subcontractors. In some areas of the country, particularly in the West, bargaining has become more regionalized, and in some segments of the industry (e.g., pipeline construction) it is national. Still, the variations on this general picture are myriad. Some 1,500 separate labor agreements are negotiated in the industry each year.

UNIONIZATION IN THE HOMEBUILDING INDUSTRY

As noted earlier, most residential construction is performed on a nonunion basis. In a few metropolitan areas, homebuilding

---

[52] Mills, *op cit.*, p. 11.

[53] Sumichrast and Frankel, *op. cit.*, p. 213.

is still heavily unionized, but even in some of these localities the unions have been gradually losing strength as new building has moved farther away from the central city. In the great majority of labor markets, single-family residences are erected almost exclusively with nonunion labor and low-rise apartment construction also tends to be nonunion. As a general proposition, the presence of unions in the residential sector is limited to apartment buildings with elevators and four or more stories. It is impossible to give precise proportions here, since published data break down residential projects by number of units rather than by height.

Although available data on union membership in residential construction are fragmentary, it is commonly acknowledged that homebuilding has historically been less organized than most other sectors. One early Bureau of Labor Statistics study found that about one-third of the workers sampled were engaged in residential work and of these less than 57 percent were union members, as compared to more than 72 percent in nonresidential construction. In larger cities with a population over 250,000, union workers constituted 62 percent of the residential work force compared to 41 percent in smaller cities.[54]

Subsequent analyses by Richard Scheuch and by William Haber and Harold M. Levinson argue that the Bureau's figures were too high, but that by 1950 union membership in the residential sector, and particularly in single-family building, had grown substantially.[55] In an interview with Scheuch in 1950, William Tobin, Industrial Relations Director for the National Association of Home Builders, estimated that some 65 percent of the workers employed by member firms were unionized. Scheuch's own estimates of the extent of organization in 25 areas throughout the country, based largely on personal interviews and congressional testimony, were that thirteen of them were almost entirely union, and six of these had changed from a dominantly nonunion character during and after the war. He noted further that even in

---

[54] The reader is reminded that residential construction includes apartment building, thus the figures for homebuilding undoubtedly would be lower. See Edward P. Sanford, "Wage Rates and Hours of Labor in the Building Trades," *Monthly Labor Review*, Vol. XXXIV (August 1937), pp. 281-300.

[55] William Haber and Harold M. Levinson, *Labor Relations and Productivity in the Building Trades* (Ann Arbor: University of Michigan, 1956), pp. 34-37; and Richard Scheuch, "The Labor Factor in Residential Construction," Ph.D. dissertation, Princeton University, 1951.

many of those areas which were nonunion in the late 1940's certain individual trades were heavily organized.

Haber and Levinson also found considerable disparity in union strength among the sixteen cities they surveyed, although their interviews with industry and union representatives in a given locality generally were consistent. The authors also noted "important variations among crafts, with the mechanical trades usually better organized than the overall average might indicate."[56] Their estimates of unionization in individual cities, along with the more qualitative appraisals by Scheuch, are presented in Table 19.

The period between World War II and the Korean Conflict probably represents the high-water mark of union representation in homebuilding for the past four decades. The years after World War II witnessed an explosive expansion in the demand for all kinds of construction, particularly residential, and the building trades unions for a sustained time had a congenial environment for their organizational efforts. Between 1954 and 1965, however, the average annual unemployment rate for construction workers never fell below 10 percent.[57] Although the picture changed for the better in the late 1960's, it had already begun to darken again by 1970.[58] Furthermore, as noted above, residential building has tended to move increasingly away from the central cities, making it progressively difficult for the unions in some areas to maintain control.

In the 1969 membership survey, the National Association of Home Builders reported that 18.6 percent of the respondents were union employers and 32.7 percent were strictly nonunion.[59] The remainder presumably used union workers or subcontractors in some trades only. In 1971, the Labor Relations Department of the NAHB asked the Executive Officers of its local affiliates to estimate the proportion of nonunion building in their areas. Although the responses were incomplete, certain scattered reports are instructive. Buffalo, for example, was estimated by both Scheuch and Haber/Levinson as solidly union; however, now there is virtually no union representation in single-family build-

---

56 Haber and Levinson, *op. cit.*, p. 36.

57 U.S. Department of Labor, *loc. cit.*

58 Even in the late 1960's, although the market in construction generally was healthy, homebuilding remained depressed largely because of the unavailability of mortgage funds.

59 Sumichrast and Frankel, *op. cit.*, p. 202.

TABLE 19. *Homebuilding Industry*
Estimates of Union Strength in Residential Construction
By Scheuch and Haber/Levinson, by Selected Areas

Area	Scheuch 1949	Haber/Levinson Percent Range 1952
Atlanta, Ga.	mostly nonunion	—
Baltimore, Md.	open shop	—
Battle Creek, Mich.	———	0-25
Boston, Mass.	strongly unionized	50-75
Buffalo, N.Y.	unionized after war	75-90
Cape Cod, Mass.	nonunion	—
Charleston, S.C.	———	75-90
Charlotte, N.C.	———	0-25
Chicago, Ill.	strongly unionized	95-100
Cincinnati, Ohio	———	0-25
Cleveland, Ohio	———	95-100
Columbus, Ohio	open shop	0-25
Dallas, Tex.	mostly nonunion	—
Denver, Colo.	unionized after war	—
Detroit, Mich.	unionized after war	75-90
Grand Rapids, Mich.	———	0-25
Hartford, Conn.	strongly unionized	—
Indianapolis, Ind.	open shop	0-25
Kalamazoo, Mich.	———	0-25
Little Rock, Ark.	mostly nonunion	—
Long Island, N.Y.	unionized after war	—
Lynn, Mass.	mostly nonunion	—
Memphis, Tenn.	30 percent all union; rest union in some crafts	—
Milwaukee, Wis.	strongly unionized	—
New York, N.Y.	strongly unionized	—
Oklahoma City, Okla.	open shop	—
Philadelphia, Pa.	strongly unionized	—
Pittsburgh, Pa.	———	25-40
Portland, Oreg.	unionized after war	—
St. Louis, Mo.	strongly unionized	95-100
San Antonio, Tex.	open shop	—
San Francisco, Calif.	strongly unionized	—
Washington, D.C.	mostly open shop; some closed shop	25-40

Source: Richard Scheuch, "The Labor Factor in Residential Construction," Ph.D. dissertation, Princeton University, 1951, Chapter V; and William Haber and Harold M. Levinson, *Labor Relations and Productivity in the Building Trades* (Ann Arbor: University of Michigan, 1956), p. 252.

ing. The case of Boston is even more illuminating. Scheuch called the city "strongly unionized"; Haber and Levinson put union strength there at 50 to 75 percent, noting that "while [Scheuch's characterization] is true for the city itself, a very large volume of homes is constructed in suburbs where union strength was reported to be much weaker." [60] By 1971, according to the NAHB survey of local Executive Officers (see Table 20), Boston was totally nonunion. We do not presently know, however, the extent to which this experience has been repeated in other larger cities.

It is certainly true, therefore, that a very substantial proportion of homes in the United States is built on a nonunion or open shop basis. The NAHB tabulation of its 1971 survey of Executive Officers covered 70 cities; 53 of them were at least 90 percent nonunion. John Riley, Labor Counsel for the NAHB, has estimated that 80 percent of all single-family and small apartment units were constructed by nonunion builders,[61] the same figure cited by Mills.[62] The President's Committee on Urban Housing concluded in 1968 that "significantly less than half of homebuilding employees are covered by collective bargaining agreements." [63] If these assessments are even roughly accurate, it would appear that there has indeed been a major decline in union organization in residential construction since the 1950's.

Future publications by the Industrial Research Unit will delve in more detail into the history and patterns of unionization in both residential and nonresidential construction; however, in regard to homebuilding, some of the salient factors behind its largely nonunion character have been identified here: the movement into suburban and nonurban areas; the small size and broad dispersion of firms; and the specialization which limits the competitive threat to union standards posed by builders. These and other factors are being explored more deeply in the research currently underway. For now, however, it may be useful to note briefly some of the implications of the relative ab-

---

[60] Haber and Levinson, *loc. cit.*

[61] Interview, April 6, 1972. Mr. Riley emphasized that the figure is only an informed guess.

[62] Mills, *op. cit.*, p. 17.

[63] See President's Committee on Urban Housing, *A Decent Home* (Washington, D.C.: Government Printing Office, 1969), p. 163.

TABLE 20. *Homebuilding Industry*
Union Strength in Single-Family Building as Estimated by NAHB Local Executive Officers, 1971

City	Percent of Single-Family Union	City	Percent of Single-Family Union	City	Percent of Single-Family Union
Birmingham, Ala.	—	Iowa City, Iowa	—	Canton, Ohio	5
Gadsden, Ala.	2	Sioux City, Iowa	—	Cleveland, Ohio	90
Phoenix, Ariz.	95	Wichita, Kans.	—	Toledo, Ohio	98
Fort Smith, Ark.	—	Lexington, Ky.	—	Ardmore, Okla.	—
Mt. Home, Ark.	—	Louisville, Ky.	—	Cresco, Pa.	5
Fresno, Calif.	3	Lafayette, La.	—	Providence, R.I.	—
Los Angeles, Calif.	100	Baltimore, Md.	—	Columbia, S.C.	—
Modesto, Calif.	—	Boston, Mass.	—	Orangeburg, S.C.	—
San Diego, Calif.	95	Davison, Mich.	20	Sioux Falls, S.D.	—
San Francisco, Calif.	100	Grand Rapids, Mich.	—	Chattanooga, Tenn.	10
Colorado Springs, Colo.	10	Kalamazoo, Mich.	—	Austin, Tex.	—
Denver, Colo.	70	Minneapolis, Minn.	97	Beaumont, Tex.	100
Gainesville, Fla.	—	St. Louis, Mo.	100	Lubbock, Tex.	—
St. Petersburg, Fla.	—	Billings, Mont.	20	Wichita Falls, Tex.	—
Tampa, Fla.	—	Merrimack, N.H.	—	Ogden, Utah	5
W. Palm Beach, Fla.	25	Dover, N.H.	5	Hampton, Va.	5
Atlanta, Ga.	2	Basking Ridge, N.J.	—	Lynchburg, Va.	—
Valdosta, Ga.	—	Binghamton, N.Y.	—	Roanoke, Va.	—
Honolulu, Hawaii	5	Buffalo, N.Y.	5	Kennewick, Wash.	50
Chicago, Ill.	100	Syracuse, N.Y.	—	Spokane, Wash.	—
Bloomington, Ind.	5	Charlotte, N.C.	—	Tacoma, Wash.	80
Elkhart, Ind.	—	Greensboro, N.C.	—	Morgantown, W.Va.	—
Ames, Iowa	—	Fargo, N.D.	10	Milwaukee, Wis.	100
				Wausau, Wis.	—

Source: 1971 NAHB survey (unpublished).

sence of unions in residential construction as they pertain to the labor market issues to be examined in subsequent chapters.

Unions in construction play a significant role in a variety of labor market activities: (1) they negotiate wages; (2) they seek to conserve work opportunities resisting encroachment on their member's work jurisdiction; (3) they administer hiring halls, and (4) they are instrumental in the governing of apprenticeship programs. The absence of unions in homebuilding affects its labor market and its firms in various ways. There are no limitations, for example, on the hiring process, and the employer is free to hire whomever he can attract on his own. At the same time, the access of both workers and employers to labor market information (that is, which workers are available and which firms have job openings) is dependent to a large extent upon an informal network of personal contacts. In the absence of a central information exchange, job data are disseminated differently and perhaps slowly and incompletely.

The union role in manpower training is only partially filled in homebuilding. As we shall see in Chapter IV, there is relatively little formal training in the industry. On the other hand, the process of informal skill development is probably eased when there are no craft demarcations to retard unskilled workers from developing a trade skill gradually. The absence of such jurisdictional restrictions also results in greater flexibility in allocating the work force among different tasks, a subject to be considered in Chapters III and IV. Indeed, many nonunion contractors report that their labor costs are lower even while paying union rates because of this greater flexibility and the absence of restrictive rules. Still, as will be seen in Chapter VI, wage rates in homebuilding tend to be substantially lower than in the union sector, and they also tend to cover a much wider range, so that individual rates are more closely tied to the productivity of the worker. Finally, although there are no hard data, it is often alleged that nonunion construction affords a greater opportunity for minorities because, it is argued, minority underrepresentation in construction is due largely to racial barriers erected by the building trades unions (see Chapter VII).

Even within the nonunion sector, there are important variations in the ways in which labor markets operate. The focus here will be on homebuilding. The differences in labor market characteristics among the various industrial subdivisions of construction will be a central element of the major study underway.

CHAPTER III

Labor Markets: The Problem of Seasonality

It is the function of any labor market to provide mechanisms by which to attract and allocate workers in response to changes in the demand for labor. In most cases, however, these changes are fairly gradual, limited mostly to vacancies caused by death, retirement, or interindustry mobility and to secular or cyclical movements in total employment. The labor market in construction carries an uncommonly heavy burden because within a brief period of time each year the need for workers rises significantly. This point is dramatically illustrated by Myers and Swerdloff:

> From its low point in February to its peak in August, contract construction—the major and most changeable part of the construction industry—adds enough workers to staff the entire motor vehicle manufacturing industry. Six months later employment will have dropped by approximately the same number.[64]

This seasonal volatility is felt in various ways: large numbers of additional workers must be found to fill peak-season manpower requirements, workers who will later be released as building activity ebbs. This rapid and temporary accretion, moreover, must produce a work force comprised of skilled craftsmen to a relatively large extent. Although certain internal adjustments may allow the absorption of a disproportionate number of unskilled workers, some of the new entrants must have specific industry-related skills. At the same time, large-scale training efforts to fill short-term seasonal needs are neither feasible nor economically justifiable.

The impact of seasonality on many construction workers is considerable. Although gross unemployment figures reflect joblessness of all kinds, there is evidence that the seasonal com-

---

[64] Robert J. Myers and Sol Swerdloff, "Seasonality and Construction," *Monthly Labor Review*, Vol. XC, no. 9 (September 1967), p. 2.
100.

ponent is predominant, accounting for perhaps 40 percent of the total which is double the national average.[65] As another measure of seasonal influence, the Bureau of Labor Statistics has compiled annual hours figures for various trades in four areas. Among workers who had 700 hours or more, discounting those with only a temporary attachment to the construction labor force, annual hours tended to fall in most cases within a range of 1,400 to 1,600, or three-fourths of a normal work year.[66] These limited work opportunities are also reflected in annual earnings; despite their high hourly rates, construction workers, by and large, do not earn more than those in many manufacturing industries.[67] At the same time, as we shall see, the problem of seasonal unemployment is somewhat less severe in homebuilding than in other sectors of construction.

Although the effect of seasonal weather conditions on the cost of construction is not easily quantified, there is little doubt that winter building does entail some additional expenses. Structures must be enclosed and heated, equipment must be moved in mud and ice, materials must be protected against freezing, and in very low temperatures workers are simply less efficient. The magnitude of the additional costs will vary by project, although it appears that there are few projects which are totally infeasible in the winter months. According to the Bureau of Labor Statistics, "Judgments obtained from cautious but knowledgeable officials put these costs in the case of building structures at not more than 5 percent."[68] Indeed, many Western nations have initiated subsidy programs to encourage winter building by defraying expense.[69]

Perhaps the major cost of seasonality is borne by everyone in the form of underutilization of human resources. Although

---

65 U.S. Bureau of Labor Statistics, *Seasonality and Manpower in Construction*, Bulletin 1642 (Washington, D.C.: Government Printing Office, 1970), p. 100.

66 *Ibid.*, p. 71.

67 U.S. Bureau of Labor Statistics, *Compensation in the Construction Industry*, Bulletin 1656 (Washington, D.C.: Government Printing Office, 1970), p. 77.

68 U.S. Bureau of Labor Statistics, *Seasonality, op. cit.* p. 29.

69 Jan Wittrock, *Reducing Seasonal Employment in the Construction Industy*, (Paris: Organization for Economic Cooperation and Development, 1967).

the character of the construction industry itself—its shifting demands, its ephemeral work groups, the difficulty of scheduling work—contributes significantly to the loss, seasonality is the major culprit. Skilled workers who are idle for months at a time constitute a substantial waste of manpower. Mills has vividly described the potential savings involved in reducing this intermittency:

> An increase of 200 hours per year in the average workyear of the present labor force would provide an additional two-thirds of a million man years of labor. To provide this volume of additional manpower to the industry without reducing intermittency would require the addition of one and one-third million persons to the industry's labor force.[70]

The years following World War II have witnessed a marked decline in seasonality compared to earlier periods. In February 1929, for example, contract construction employment stood at 70 percent of the annual average that year, rising to 124 percent in August.[71] In 1947, by contrast, the comparable figures were 85 percent and 111 percent. These ratios, however, did not change significantly for the following two decades. In very recent years, there appears to be a slightly further decline in seasonality. Table 21 traces the industry's experience since 1965. Although these gross employment figures do not distinguish homebuilding from other contract construction, an attempt shall be made to describe the seasonal patterns in residential construction specifically.

The seasonal expansion of employment remains substantial. Between February and August 1973, the industry added nearly 800,000 workers, or 25 percent of the February level. Moreover, seasonality is not uniform throughout the industry or the nation. In 1968, for example, the August-February employment ratio was 116 for general building contractors, 156 for heavy construction, 198 for highway construction, and 114 for special trades contractors. Occupationally, the ratios were 120 for carpenters, 113 for other craftsmen, and 156 for laborers. Perhaps most important, there are wide variations geographically in the incidence of seasonality, which are masked by national statistics. Since labor markets are predominantly local, the problem of labor force adjustment must be measured by the

---

[70] Mills, *op. cit.*, p. 120.

[71] U.S. Bureau of Labor Statistics, *Seasonality, op. cit.*, p. 32.

TABLE 21. *Construction Industry Number and Percent of Annual Average of Contract Construction Employment United States, February and August, 1965-1973* (Numbers in thousands)

Year	Annual Average	February		August	
		Number	Percent	Number	Percent
1965	3,186	2,691	84.5	3,546	111.3
1966	3,275	2,822	86.2	3,628	110.8
1967	3,208	2,824	88.0	3,526	109.9
1968	3,285	2,909	88.6	3,581	109.0
1969	3,473	3,021	87.0	3,731	107.4
1970	3,345	3,071	91.8	3,606	107.8
1971	3,411	2,930	85.9	3,691	108.2
1972	3,521	3,096	87.9	3,838	109.0
1973	n.a.	3,184	n.a.	3,981	n.a.

Source: U.S. Bureau of Labor Statistics, *Employment and Earnings*, various issues.

local experience. In 1967, the August-February employment ratio ranged from 111 in the West South Central states and 115 in the South Atlantic states to 149 and 145 in the West North Central and New England states.[72] Recent experience in the Buffalo SMSA is presented in Table 22.[73]

The figures in Table 22 attest to the burden placed on a local labor market by the seasonal expansion of building activity. In the relatively short span of six months, the Buffalo work force usually grows by some 40 percent. Mobility may to some extent provide workers, but this solution is not feasible when there are few unemployed workers in other localities or industries, or where workers are tied to their jobs and communities for personal reasons.[74] Thus seasonal fluctuations

[72] All 1967 and 1968 figures in this paragraph are calculated from U.S. Bureau of Labor Statistics data; *Ibid.*, pp. 31-33.

[73] Since figures for a single month in a limited area may be subject to considerable random influence, quarterly averages are used instead.

[74] For a discussion of "the limitations of mobility," see Mills, *op. cit.*, pp. 137-138.

TABLE 22. *Construction Industry*
First-Quarter and Third-Quarter Employment
In Contract Construction
Buffalo SMSA, 1966-1972
(Numbers in thousands)

Year	First Quarter	Third Quarter	Percent Change
1967	16.7	23.3	39.5
1968	16.2	24.3	50.0
1969	17.2	21.8 [a]	26.7
1970	15.9	21.9	37.7
1971	14.6	20.0	37.0
1972	13.3	18.1 [a]	36.1

Source: New York State Division of Employment, *Employment Review*, various issues.

[a] Figures affected by midsummer strikes in 1969 and 1972.

require other kinds of labor market adjustments, the mechanics of which are complex and incompletely understood.

SEASONALITY AND THE HOMEBUILDING INDUSTRY

It is very difficult to analyze the nature and impact of seasonality in residential construction. The major stumbling block, of course, is the absence of data which will distinguish homebuilding from the rest of construction. Although it is generally known that building construction is less seasonal than heavy and highway construction, and that general contractors experience more volatility than special trades contractors, it is still difficult to make a firm judgment as to the relative instability of homebuilding and other sectors.

There are some statistical series relating to both the product market and the labor market which provide monthly data on various indexes of activity. The two major series on the product market side are housing permits and housing starts. Since permits may be taken out well in advance of actual construction, housing starts provide a better measure of seasonal movements. Tables 23 and 24 give a percentage distribution of starts in each quarter for the years 1963 through 1972,

TABLE 23. *Homebuilding Industry*
Percent Distribution of Housing Starts, by Number of Units Per Quarter
United States, 1963-1972

Year	Total				Single-Unit				Multiunit			
	I	II	III	IV	I	II	III	IV	I	II	III	IV
1963	18.3	30.4	27.4	23.9	17.5	31.8	28.5	22.3	19.7	28.3	25.5	26.5
1964	21.2	29.8	26.1	22.9	20.7	29.8	26.8	22.6	22.0	29.8	24.7	23.4
1965	19.4	31.0	26.5	23.2	18.8	30.8	27.6	22.7	20.4	31.3	24.4	23.9
1966	23.7	33.5	24.7	18.1	23.1	33.2	25.4	18.2	24.7	34.0	23.4	17.9
1967	16.5	28.9	28.9	25.7	17.5	30.2	28.9	23.4	14.7	26.6	28.9	29.8
1968	19.3	29.2	27.4	24.1	20.0	29.6	27.7	22.7	18.4	28.7	26.9	26.0
1969	22.4	31.2	25.8	20.6	21.1	32.0	26.5	20.5	23.9	30.3	25.0	20.7
1970	18.0	27.2	27.8	27.0	16.8	28.5	28.2	26.5	19.5	25.6	27.4	27.6
1971	18.6	29.0	27.8	24.6	17.8	30.3	27.9	24.0	19.7	27.3	27.5	25.4
1972	21.4	28.0	27.1	23.5	20.2	29.6	28.3	22.0	23.0	26.1	25.5	24.7
Average	19.9	29.8	27.0	23.4	19.4	30.6	27.6	22.5	20.6	28.8	25.9	24.7

Source: U.S. Department of Commerce, Business and Defense Services Administration, *Construction Review*, various issues.

TABLE 24. *Homebuilding Industry*
Percent Distribution of Housing Starts, by Region Per Quarter
United States, 1963-1972

Year	Northeast				North Central				South				West			
	I	II	III	IV	I	II	III	IV	I	II	III	IV	I	II	III	IV
1963	12.6	34.8	28.2	24.4	10.2	35.9	29.1	24.7	21.6	27.0	27.8	23.6	23.6	28.3	24.8	23.3
1964	12.7	31.8	28.0	27.5	16.6	32.8	27.5	23.1	23.6	28.2	25.1	23.1	27.8	28.1	24.8	19.3
1965	13.9	34.2	27.0	25.0	12.8	33.6	28.4	25.1	23.1	28.4	26.3	22.3	26.2	29.6	23.6	20.6
1966	17.9	35.3	28.3	18.5	19.5	36.6	26.2	17.8	26.3	31.1	23.7	18.9	29.8	32.6	21.3	16.3
1967	12.2	29.1	32.7	26.0	12.5	31.8	29.0	26.8	20.1	27.6	27.0	25.3	18.3	27.2	29.6	24.9
1968	12.3	31.2	34.7	21.8	15.9	32.1	27.9	24.1	22.1	26.9	25.8	25.1	23.4	28.9	24.1	23.6
1969	18.4	38.5	23.9	19.2	21.3	31.7	28.2	18.8	25.4	28.6	25.1	20.9	20.7	30.8	25.6	22.9
1970	15.8	28.8	30.0	25.4	14.6	29.6	27.7	28.2	19.2	26.4	27.2	27.2	20.4	25.4	27.7	26.4
1971	13.0	29.9	30.6	26.6	14.0	30.6	29.9	25.5	20.6	27.9	26.7	24.7	22.3	28.9	26.2	22.6
1972	15.4	27.2	29.9	27.5	15.6	31.6	30.5	22.2	22.9	27.5	25.5	24.0	27.2	26.5	25.5	20.8
Average	14.4	32.1	29.3	24.2	15.3	32.6	28.4	23.6	22.5	28.0	26.0	23.5	24.0	28.6	25.3	22.1

Source: U.S. Department of Commerce, Business and Defense Services Administration, *Construction Review*, various issues.

distinguishing single-unit and multiunit dwellings, as well as showing variations by geographic area. These figures depict a pronounced seasonal pattern in residential building. The peak of housing starts in the second quarter of the year should not be confused with the peak of total employment which generally occurs in the third quarter. Not surprisingly, seasonality appears to be greater in single-family homes than in multiunit dwellings, since technological methods and planning to facilitate winter building are more easily and economically adapted to larger projects. The relatively stable patterns in the South and West, as opposed to the Northeast and North Central states, are also to be expected.

The patterns in Tables 23 and 24, of course, are affected by cyclical factors, as well as seasonal influences. A somewhat more accurate reading of seasonality may be derived from a series of seasonally adjusted data on housing starts published annually in *Construction Review*. Seasonal indexes are calculated by dividing actual monthly starts by the adjusted figure; these seasonal factors for the years 1966 through 1971 are shown in Table 25. Again the seasonal movement is readily discernible. The differences among areas also show up in the seasonal factors, as the figures for 1971 in the four major regions demonstrate (Table 26).

Unfortunately, data on building activity are not strictly comparable with such labor market variables as employment and unemployment.[75] Furthermore, as mentioned earlier, there is no efficient way to extract employment figures on residential construction from data series relating to the industry as a whole. In short, one cannot make a firm judgment on whether employment in homebuilding is more or less stable than that in other sectors. This is, as we shall see, of more than academic concern, since many of the residential employers in Erie County attribute the differences between union and nonunion wage rates to the relatively steady character of work in nonunion homebuilding as opposed to those sectors of nonresidential construction which are strongly unionized.

We may, however, seek a hint as to the relative seasonality of residential construction by examining employment patterns in forms of the two industrial divisions designated "subdividers

[75] U.S. Department of Commerce, "Noncomparability of Statistics on Contract Construction Employment and Value of Construction Put-in-Place," *Construction Review*, Vol. I, no. 1 (March 1955).

TABLE 25. *Homebuilding Industry*
Seasonal Indexes of Private Housing Starts
United States, 1968-1972

Month	1968	1969	1970	1971	1972
January	.700	.689	.719	.733	.742
February	.668	.634	.674	.684	.726
March	1.037	1.014	1.009	1.040	1.067
April	1.251	1.252	1.253	1.237	1.161
May	1.201	1.179	1.172	1.164	1.180
June	1.179	1.157	1.163	1.158	1.173
July	1.110	1.098	1.122	1.115	1.110
August	1.096	1.103	1.102	1.106	1.144
September	1.035	1.029	1.026	1.028	1.011
October	1.076	1.072	1.064	1.058	1.071
November	.936	.924	.939	.935	.937
December	.775	.759	.749	.743	.774

Source: U.S. Department of Commerce, Business and Defense Services Administration, *Construction Review*, Table B-1, various years.

and developers" and "operative builders," whose official classification falls in the real estate industry. Although these firms constitute only a fraction of all builders, they are the only ones that are fairly clearly distinguished from nonresidential construction employers in government employment data series. There is a clear seasonal pattern in the two groups, generally reaching a low in January and a high in July as opposed to February and August for most of construction. Table 27 presents employment levels for the appropriate months in the two residential groupings and, for comparison, in general building construction. In evaluating the table, the reader should be aware of at least three limitations: (1) general building construction includes many residential contractors; (2) the data for general contractors refer to "production workers," although the other series include all male workers; and (3) the employees of subcontractors are excluded.

Finally, a Bureau of Labor Statistics survey of over 4,200 residential contractors and subcontractors elicited information

TABLE 26. *Homebuilding Industry*
Seasonal Indexes of Private Housing Starts
By Region, 1971

Month	Northeast	North Central	South	West
January	.480	.500	.780	.957
February	.421	.442	.795	.809
March	.819	.994	1.118	1.055
April	1.420	1.468	1.106	1.171
May	1.363	1.121	1.134	1.151
June	1.128	1.200	1.165	1.094
July	1.421	1.115	1.064	1.043
August	1.200	1.198	1.093	.991
September	1.027	1.060	1.003	1.048
October	1.066	1.137	.004	1.017
November	.988	1.034	.906	.866
December	.634	.676	.786	.818

Source: U.S. Department of Commerce, Business and Defense Services Administration, *Construction Review*, 1971, Table B-1, various issues.

on employment patterns during 1969.[76] Between January and July, total employment among those contractors rose by 22.5 percent. Not surprisingly, the increases were greatest among contractors engaged in "basic" work (excavating and grading, masonry, carpentry, and cement) and least among "finishing" contractors (plastering, tile, and flooring). It appears, however, that the Bureau of Labor Statistics sample is not representative of residential employers as a group, since employment in July 1969 averaged about 19 workers per contractor[77] which is substantially higher than the industry-wide averages noted in Chapter II. Since larger contractors are more able to utilize technological and organizational techniques to operate through the winter, the January-July 22.5 percent increase is probably a

[76] U.S. Bureau of Labor Statistics, *Labor and Material Requirements for Construction of Private Single-Family Houses*, Bulletin 1755 (Washington, D.C.: Government Printing Office, 1972), p. 21.

[77] Total employment among the over 4,200 contractors was 78,664. *Ibid.*

TABLE 27. *Construction Industry*
Employment During High and Low Months
By General Building Contractors and Subdividers, Developers, and Operative Builders
United States, February and August, 1967-1972
(Numbers in thousands)

	All Employers			General Building Contractors [a]			Subdividers, Developers, & Operative Builders [b]		
Year	February	August	Percent Change	February	August	Percent Change	February	August	Percent Change
1967	891.0	1,067.4	19.8	749.9	921.6	22.9	77.0	104.0	35.1
1968	911.9	1,078.6	18.3	763.1	925.2	21.2	86.5	108.4	25.3
1969	950.9	1,137.9	9.1	795.5	974.4	22.5	95.2	121.0	27.1
1970	938.8	n.a.	n.a.	777.7	876.4	12.7	96.0	119.2	24.2
1971	827.8	1,011.9	22.2	668.8	846.1	26.5	94.4 [c]	137.9 [c]	46.1
1972	841.9	1,120.1	33.0	677.0	931.0	37.5	120.3 [c]	168.6 [c]	40.1

Source: U.S. Bureau of Labor Statistics, *Employment and Earnings, United States, 1969-70*; Bulletin 1312-8 (1967-70), (Washington, D.C.: Government Printing Office).

[a] Data are for construction workers only.

[b] Data are for male employees only.

[c] Data are for January and July, figures on female employment not available for February and August.

very conservative estimate of seasonal growth for homebuilding as a whole.

The data presented in the foregoing pages suggest that the magnitude of seasonal fluctuations in employment is of a comparable order in both residential construction and other building construction, but perhaps somewhat greater in residential.[78] The problem of seasonality in Erie County, and employers' response to it, is examined in the following section. As we shall see, the methods of labor force adjustment may provide some insights into the reasons why employers feel that homebuilding is more stable than other construction work and the extent to which different seasonal patterns influence wage differentials within the industry.

Seasonality in Erie County

Seasonal fluctuations in Erie County tend, for obvious reasons, to be more severe than in the nation as a whole. Contract construction employment in the Buffalo-area SMSA (which includes Niagara County as well) typically rises by 60 to 70 percent between February and August.[79] Although the increase for residential buildings alone is undoubtedly smaller (the aggregate figures include the exceptionally seasonal heavy and highway sector), the seasonal movement is nevertheless marked. If the previous discussion is any guide, there is no reason to believe that the residential swings are substantially less than those in building construction generally.

It is, therefore, somewhat paradoxical that many employers interviewed in Erie County alluded to the relative stability of residential work. When asked why some workers continue to work in homebuilding rather than moving into [higher-paying] branches of the industry, well over one-half of the respondents made some mention of steady work. The comments below are illustrative.

—Though they pay a higher rate, they work fewer hours a year, whereas in homebuilding you're paid at a lower rate but can work more hours. So you wind up with more income.

---

[78] Mills states that homebuilding is more seasonal than the rest of construction, although he offers no evidence. Mills, *op. cit.*, p. 121.

[79] See various recent issues of *Employment Review*, published by the New York State Division of Employment.

—In our company, we try to counteract this by keeping our people working all year. Our contention is that lower hourly pay makes up for working only half a year.

—[It is] due to the consistency of work in homebuilding industry, as compared to the inconsistency of the construction industry.

—Because I let them have full-time employment. They know who they'll be working for from year to year.

Despite the steadiness of employment asserted by many of the respondents, it was apparent that fluctuations in employment over a twelve-month period are marked. Respondents were asked to report the highest number and the lowest number of on-site employees on their payroll during the previous year. In only 12 cases (14 percent) were the two figures identical, and most of these were small firms with one or two workers. About one-half the employers indicated that their highest employment level was at least double their lowest level. The total high and low employment figures for each type of employer are shown in Table 28. Some of the categories in this table are greatly influenced by the experience of a single employer. Thus most of the differences for electrical contractors are accounted for by one firm with a high of 15 workers and a low of seven. Similarly, one plumbing contractor reported a high of 25 workers and a low of seven. On the other hand, the large differences for carpentry, drywall, and roofing do reflect the general experience of those firms contacted.

To be sure, a firm's employment in the course of a year will fluctuate for reasons unrelated to seasonal factors. At the same time, there are other indications that seasonality does play a prominent role. For example, respondents were asked whether the difference between their high and low employment levels was at all attributable to "temporary people you hired to help out during a busy period." About 60 percent of those interviewed responded in the affirmative. The total number of temporary workers reported by those who answered affirmatively was 221, which constitutes about one-half the aggregate high-low difference.

For the most part, these workers were regarded as general laborers or helpers. When asked what kinds of jobs the temporary workers were given to do, employers generally responded in such terms as "cleaning up," "hauling and lifting," "pick up material," "menial jobs," and "nothing that required an awful lot of thought." When asked to characterize the skill

TABLE 28. *Homebuilding Industry*
High and Low Employment Levels for Employers
For 12-Month Period, by Type of Employer
Erie County, New York, 1971

Type of Employer	High	Low
Builder	182	121
Subcontractor:		
Carpentry	250	88
Cement	109	66
Drywall	105	40
Electrical	31	21
Excavating	53	30
Heating	79	57
Painting	49	35
Plumbing	95	53
Roofing	52	26
Subcontractor total	823	416

Source: Erie County survey, 1971.

level of temporary workers, 27 said unskilled, 18 said somewhat skilled, and only 6 said highly skilled.

Most of the temporary workers were students. Of the 52 employers who reported hiring temporary help, 43 indicated that some or all of these workers were students. Four respondents had hired high school teachers, and 15 others had taken on workers who had other full-time jobs. The primary occupations of these "moonlighters" were varied, including a steel worker, farmer, policeman, and fireman.

The hiring of temporary workers, as well as such other responses to temporary manpower needs as overtime, work by the employer himself, or subcontracting, are necessary expedients when requirements expand during the peak season. When demand falls, other employment adjustments occur, and these are not limited simply to the release of students and other temporary employees. For example, the composition of work performed by regular employees may change. Employers in

Erie County were asked the following question: "Are any of your regular full-time workers assigned different kinds of work in the summer and winter?" Nineteen respondents answered in the affirmative. The changes in work assignments took two forms: (1) craftsmen were given work in the winter which would normally be performed by unskilled labor; and (2) journeymen were assigned only those kinds of work in the summer which could not be performed in cold weather, leaving other work for the winter months. Some specific examples are:

—If work is slow in winter we would not keep on the helpers. We have the carpenters perhaps carry their own two-by-fours during winter months.

—They would do painting and drywall in winter, carpentry in summer.

—They'll do the bulk work rather than have the helpers do the bulk work.

—Our concrete finishers and laborers do carpentry work in the winter time.

—Snow plowing in winter, truckdriving and bulldozing in summer.

—Truck driver might repair truck during winter.

Some employers find other ways to keep their workers busy during the slow season: (1) houses may be built on speculation; (2) if demand for homes is slack, firms may construct apartments or workers may do maintenance on existing apartments; (3) the opportunity may be used to overhaul equipment. The practice of holding on to workers despite a limited need for them is widespread because employers are generally fearful of losing their key people. Thus, when asked if they ever kept workers on the payroll in the off-season even if there was little work for them to do, about 60 percent of the Erie County respondents answered in the affirmative.

Another kind of adjustment, of course, is to lay workers off until building activity resumes. Yet when asked whether any of their regular employees typically experienced layoffs during the winter, two-thirds of the responses were negative. Most of the affirmative responses, moreover, came from those areas where work is particularly vulnerable to weather conditions, such as cement pouring and excavation.

This description of the various adjustments to seasonal needs may provide some explanation of the inconsistency between volatile employment patterns and the frequent allusion by employers to the relative stability of work in homebuilding. First, however, it is necessary to understand that the labor force in construction generally is comprised of at least three identifiable groups of workers, and that the relative size of each group is different not only among crafts, but also among sectors. These three groups have been described as follows:

> A proportion of the labor force (this varies, of course, by craft), which probably amounts to at least a quarter of the total, has substantially full opportunities for annual employment. This group [Group A] consists of workers employed by contractors on a year-round basis, those who are employed on loan-duration jobs, those in shortage areas, and very experienced individual workers in sufficient demand so that opportunities for year-round employment are in fact offered to them.
>
> A second group of workers [Group B]—at least half the total—who carry the brunt of the impact of seasonal unemployment, has substantially less than full-time or full-year earnings opportunity, but relies upon the industry for virtually all earnings opportunity.
>
> The third group [Group C] of workers in the industry's work force is in and out of the industry. It is in part a work force that is added during the summer months as college students and teachers become available. It is in part a work force drawn from other industries and from farming, particularly during periods of peak construction employment.[80]

Although it is one of general tendencies and not applicable to every worker in the industry the pattern of employment in the residential construction industry can be further described. During the winter months, employment is limited largely to those workers in the first group described above. These workers frequently develop an attachment to a particular employer and are those whom the employer is reluctant to dismiss for fear of losing them permanently. In many cases, the employment experience of these workers is much the same as that in a manufacturing facility.

As the need for workers expands, members of the second group become reemployed, although frequently not with the same employer who employed them at first. When this pool is depleted, employers then turn to the third group. As noted, the difference between groups B and C is that the former exhibits a more or less permanent attachment to the industry (not necessarily to an individual employer), whereas the latter's attachment is more typically ephemeral and secondary. Group B, in other words, is generally unemployed during the off-season and Group C is generally preoccupied with different ac-

---

[80] "Report by the Secretaries of Labor and Commerce on Seasonality of Employment in the Construction Industry," *Daily Labor Report* (October 8, 1968), p. E-2.

tivities. Furthermore, workers in Group B will ordinarily have a higher level of skill on the average than those in Group C. As demand for labor passes its peak, workers in Group C and then in Group B are released, until the work force once again is comprised only of the year-round core.

The results of this inquiry suggest that the relative sizes of Groups A, B, and C may be different in homebuilding from those in other types of construction. Specifically, it appears that in homebuilding Groups A and C are proportionally larger and Group B is smaller. In other words, a smaller percentage of workers *who are attached to residential construction* follow the pattern of frequent moves from employer to employer with lengthy spells of joblessness in the winter months. As employment expands seasonally, relatively more of the new workers are temporary additions to the work force (Group C) and relatively fewer are those called back from layoff (Group B). When employment contracts, the released workers return to their primary activity rather than to the unemployment rolls.

The foregoing analysis suggests why employers tend to characterize residential work as steady, although aggregate employment patterns are similar to those in construction generally. The workers who, according to the earlier quotation, "carry the brunt of the impact of seasonal unemployment" constitute a smaller share of the work force in housing construction than in other sectors of the industry. In short, although seasonal *employment* patterns may appear the same, seasonal *unemployment* is indeed less severe. Furthermore, since there is apparently less job mobility in homebuilding, there is perforce less frictional unemployment, thus lending additional credence to the idea that employment is more stable in residential construction.

ADJUSTMENTS TO SEASONAL FLUCTUATIONS IN CONSTRUCTION

There is, unfortunately, no way at present to test rigorously the extent to and ways in which seasonal adjustments in homebuilding differ from those in the rest of the industry. To do so would require comprehensive data by sector on annual hours worked and patterns of unemployment. These data do not exist. It is important to note, however, that many of the adjustments described in the previous section are not uncommon in nonresidential construction.[81] The critical question is the relative importance of the various adjustment techniques in different parts of the industry.

This question is important because it relates to the efficiency with which a problem endemic to the industry is confronted. Obviously a process which results in relatively little waste of human resources is preferable to one which involves lengthy spells of idleness. Given a certain amount of volatility in manpower needs within a twelve-month period, labor force adjustments may occur along a spectrum. At one extreme there is a stable work force which is progressively supplemented by temporary workers who return to other kinds of activity when they are no longer needed. To the extent that this process works smoothly, and that the temporary workers have the necessary skill to perform at least some type of construction work, it appears to be an optimal response to changing requirements.

At the other extreme of the spectrum, the basic work force may include enough workers to fill peak-season needs. As these workers are released in the fall and winter months, they become unemployed and their productive capacity is entirely wasted. Although neither residential nor nonresidential construction is described by either of these extremes, the evidence tentatively points to the conclusion that homebuilding rests closer to the optimum than do other sectors of construction. The remainder of this section will discuss adjustment patterns in nonresidential construction for the purpose of highlighting both its similarities with and differences from homebuilding. As will be seen, the relative flexibility with which homebuilders are able both to expand and to utilize their work force is due in no small measure to the absence of unions.

Expansion of the Labor Force

In construction as a whole, the unemployment rolls constitute a significant source of peak-season manpower. As the weather warms and more contracts are let, unemployed construction workers return to work. By August of any given year the number of unemployed whose last reported job was in construction will have fallen to less than one-half of the mid-

[81] See, for example, Foster, "Labor Force Adjustments," *op. cit.*, pp. 528-540

TABLE 29. *Construction Industry*
Employment and Unemployment in Contract Construction[a]
United States, February and August, 1967-1972

	Employment			Unemployment		
Year	February	August	Change	February	August	Change
1967	2,824	3,526	702	420	162	258
1968	2,909	3,581	672	421	164	257
1969	3,021	3,731	710	336	181	155
1970	3,070	3,651	581	486	325	161
1971	2,930	3,691	761	669	299	370
1972	3,096	3,838	742	687	369	318

Source: U.S. Bureau of Labor Statistics, *Employment and Earnings, Statistics for the United States, 1909-72,* Bulletin No. 1312-8 (Washington, D.C.: Government Printing Office).

[a] Figures are not entirely comparable because employment data are based on establishment reports, whereas unemployment data are from the household survey.

winter high. At the same time, however, this reduction in unemployment accounts for only a fraction of the total increase in employment. As indicated in Table 29, in no recent year has the decline in unemployment between February and August been as much as one-half the employment rise.

Construction workers who secure other work in the off-season would not, of course, show up in the construction unemployment figures, but there is some evidence to suggest that their number is small. In 1964, Myers and Swerdloff reported that workers who were employed in construction during all four quarters (which again limits the population to those with a firm attachment to the industry) averaged only $232 of earnings in other industries, or only 3.5 percent of total earnings.[82] In the author's earlier study of construction workers in Syracuse, moreover, 716 of 748 respondents (95.7 percent) reported *no* outside employment in 1967.[83] In short, most seasonal expansion reflects an increase in the construction labor force (employment *plus* unemployment), and this increase is oc-

[82] Myers and Swerdloff, *op. cit.*, p. 4.

[83] Foster, "Labor Force Adjustments," *op. cit.*, p. 530.

casioned almost entirely by an influx of persons whose primary activity is outside the construction industry.

Much of the seasonal growth of the construction labor force shows up in government statistics in the skilled worker category (see Table 30). At the same time, it can be argued that many of the added "craftsmen" perform work requiring considerably less skill than is expected of the typical full-year workers. The highly skilled mechanical trades (electricians and plumbers) expand very little in summer, and even this expansion is almost exclusively attributable to the influx of students and other temporary workers who perform relatively menial tasks. Crafts such as cement masons and painters are extremely seasonal, and yet one hears few complaints of shortages for these relatively low-skilled crafts.[84] (In homebuilding, the reader will recall that, according to employers in Erie County, most summer workers are used for relatively unskilled tasks.)

Further indication of the dominance of low-skilled workers in the summer craftsmen group was provided in the interviews with contractors. Virtually all employers voiced some dissatisfaction with the quality of "permit men" and other temporary workers. It was very often necessary, they said, to arrange job assignments so that skilled workers were not obliged to perform tasks which could be managed by a less talented worker. This was accomplished in two ways. First, the summer craftsman might be given the easier jobs within his trade. Some operations, however, are not readily separable. Then, the answer is to introduce a partner system where the new entrant becomes essentially a helper, working under the close supervision of a skilled journeyman and following his instructions. This procedure is very common among carpenters, although other crafts such as reinforcing ironworkers and painters also may utilize it. The work of electricians and plumbers can be divided similarly, although unions in these crafts generally are able to provide relatively large numbers of apprentices for the purpose. In summary, a good deal of the demand for "summer

---

[84] For an enlightening discussion of relative skill requirements for various crafts, see George Strauss, "Apprenticeship: An Evaluation of the Need" in Arthur M. Ross (ed.), *Employment Policy and the Labor Market* (Berkeley: University of California Press, 1965), *passim* but especially pp. 308-311.

TABLE 30. *Construction Industry*
Employment and Unemployment of Craftsmen and Laborers in Contract Construction[a]
United States, February and August, 1967-1972

Year	Craftsmen[b]						Laborers					
	Employment			Unemployment			Employment			Unemployment		
	Feb.	Aug.	Change	Feb.	Aug.	Change	Feb.	Aug.	Change	Feb.	Aug.	Change
1967	2,552	2,952	400	240	73	167	544	864	320	138	71	67
1968	2,615	3,006	391	236	88	148	590	921	331	137	68	69
1969	2,661	2,997	336	173	77	96	651	969	318	110	76	34
1970	2,520	2,929	409	238	178	60	684	1,026	342	151	117	34
1971	2,735	3,250	515	344	184	160	595	1,030	435	246	109	137
1972	2,903	3,604	701	370	191	179	704	980	276	246	151	95

Source: U.S. Bureau of Labor Statistics, *Employment and Earnings*, March and September issues for each year, Tables A-10 and A-18 or A-16.

[a] These figures are not actually limited to workers in the construction *industry*. Nevertheless, it is safe to assume that most of the February to August *change* occurs in the construction industry, given its extraordinary seasonal volatility.

[b] Craftsmen equals carpenters and other construction workers.

craftsmen" can be filled by workers with only the most basic training for specific kinds of construction work.

Probably the most numerous of the new entrants are full-time students who work at construction during their summer vacation. Most students are employed as laborers, but many others secure their jobs through one of the craft unions. Indeed, most of the business agents interviewed in the author's earlier study indicated some use of student workers during peak seasons. It appears safe to conclude, however, that most students, even when nominally under the auspices of a "craft" union, do not possess the skills of a full-fledged craftsman. They generally do not receive journeyman's wages but rather the first-year apprentice's scale.[85] All the agents, moreover, indicated that the students are used primarily for "dirty work." Accordingly, it seems safe to conclude that this group of summer craftsmen is constituted primarily of what a homebuilder might term "helpers."

Upgrading

One might question, at first, whether upgrading rightfully can be considered a source of manpower, since the laborer and operative are construction workers to begin with. The answer, of course, is that every laborer who works as a craftsman vacates a job which is then filled by someone else. When asked to specify their summer manpower sources, many of the contractors and some business agents referred (without prompting) to the utilization of unskilled or semiskilled workers in order to meet a temporary need. Sometimes such upgrading is formalized, as when an individual changes his union affiliation. More often, however, the process is informal, and business agents will seldom object as long as all their members are at work. This process, of course, is very common in nonunion homebuilding as well.

Upgrading can take either of two forms. One is for the employer to use lower-skilled men from another craft; the other is to advance workers within a single craft. Since in union construction the former method may encounter serious jurisdictional barriers, the latter is probably more common where unions are involved.

---

[85] The apprentice wage is usually about 50 or 60 percent of the journeyman scale. That level is a good deal higher than what a residential contractor has to pay for a temporary helper or laborer.

The primary example of upgrading within a single union jurisdiction is the use of apprentices. During the off-season, when skilled workers are relatively plentiful, employers tend to keep apprentices, particularly those in their early years of training, on so-called "dirty work." They carry materials, fetch tools, build forms, and perform other easily learned tasks. As the supply of skilled craftsmen dwindles, the apprentice may be pressed into service to apply the knowledge he has gained through observation and helping journeymen. The menial jobs are then left to younger apprentices and to any temporary workers the employer is able to find. (In homebuilding, those workers appropriately classified as "helpers" may undergo the same experience, leaving their menial tasks to laborers.) Thus there may develop a reverse bumping all along the line: older apprentices undertake journeymen's work; younger apprentices fill in for the older ones; and the younger ones are themselves replaced by "summer craftsmen."

The same type of upgrading may also place at the journeyman level. As in any work situation, some individuals are better than others. When the market is loose, only the higher-skilled workers perform the difficult jobs. As activity increases, employers are forced to use less talented workers, although they may be slower and less efficient, leaving their old jobs to new entrants or first-year apprentices.

The second type of upgrading, the one involving a crossing of jurisdictional lines, may be less common than the kind just discussed, but it nevertheless occurs. Interviewing a business agent for the Cement Masons, the writer was told that on some jobs laborers were used extensively for cement finishing. Questioned about the jurisdictional problem, the agent replied that "all the basic trades do it." As responses from workers revealed, a significant number of carpenters, operating engineers, and even bricklayers start out as construction laborers, and it is difficult to imagine how they become craftsmen without some prior opportunity actually to perform craftsmen's work. Although some familiarity can be gained simply through observation, practical experience obviously is necessary to learn a trade. Equally obvious, unskilled workers are most likely to gain that experience when skilled labor is scarce.

Interunion upgrading is made possible by the fact that laborers often work with most of the basic trades and are thus in a position to "get a feel" for the tasks performed by craftsmen. Laborers help to pour the concrete finished by cement masons, strip the forms built by carpenters, and mix the mortar used by bricklayers. Similarly, laborers and truck drivers are at times even able to pick up the ability to operate the more simple types of construction machinery, such as bulldozers. In homebuilding, of course, the transition is made much easier by the absence of union jurisdictional rules.

In summary, the adjustment process in nonresidential construction appears more to involve idle manpower than that in homebuilding. In both sectors, however, peak-season needs are also met by the absorption of temporary workers from outside the construction work force. This absorption may operate somewhat more smoothly in homebuilding because of the absence of jurisdictional barriers and greater flexibility in setting wages and allocating manpower.

CONCLUDING REMARKS

It must be acknowledged that the discussion in this chapter of the mechanics of seasonal labor force adjustments has relied to an uncomfortable extent on limited and circumstantial evidence. In the absence of comprehensive and detailed labor market data, particularly in regard to the homebuilding industry, hard conclusions have often had to give way to speculation and inference. With these constraints in mind, and pending more systematic investigation into the labor market behavior of the various groups of construction workers, we may venture a cautious and tentative evaluation of the relative efficiency of these adjustments in the residential and nonresidential sectors.

It seems fairly safe to conclude, from the evidence assembled earlier, that seasonality is at least as much a problem in the homebuilding industry as in the construction industry on the whole. Thus far, however, the issue has been treated as merely one of generating a labor force sufficient to fulfill peak-season needs and accepting the idleness of many workers during the off-season as an unfortunate but unavoidable by-product. Yet the technical and organizational know-how to sustain at least some of the seasonal swing is presently available, and there are a number of steps which the industry could take to stabilize its production patterns.[86]

[86] See the discussion in Mills, *op.cit.*, pp. 129-132.

In the present discussion, however, we have taken the problem of seasonality as a given. In that context, our limited and uneven evidence suggests that the patterns of labor force adjustment in residential homebuilding involve less waste and more efficient utilization of manpower than do those in the unionized sectors. There appear to be fewer incidences of joblessness among the regularly attached workers who are to a greater or lesser degree skilled in construction work. In all but the most seasonally vulnerable operations, employers assert that seasonal layoffs are not routine, a condition certainly not true in the union sector.

The practice of adding temporary workers to meet peak requirements is, of course, common in both sectors, but our findings here tend to indicate that it is relatively more extensive in the homebuilding industry. To a large extent, this is a reflection of the greater flexibility enjoyed by nonunion employers both in assigning work and in setting wage rates. The union contractor who hires a marginally skilled carpenter must pay him the journeyman's wage and may not use him for tasks which would infringe upon another union's jurisdiction. The nonunion employer, on the other hand, can pay a wage commensurate with the worker's ability and can use him, for example, as a carpenter's helper, even if some of the functions would, in a union situation, be the laborer's work. Thus the union employer is more limited in his capacity to take on unskilled or semiskilled workers, and in any event it costs him more.

Other provisions in union agreements discourage the use of temporary workers. Some homebuilders continue to employ students during the school year for portions of a day or during weekends. Under union contracts, such students would frequently have to be paid for a minimum number of hours per day as well as double time on weekends. The nonunion builder is not burdened by these premiums.

In summary, these considerations suggest that the homebuilding labor market has been better able to adapt to seasonal instability. This conclusion, as noted earlier, has certain implications for some of the issues to be explored ahead. For example, in regard to wages, the analysis here indicates that there is some truth to the claim that lower rates in the homebuilding industry are in part attributable to less severe seasonal unemployment among permanently attached workers. In theory, it is difficult to dispute the idea that the market will establish a wage premium to compensate for the uncertainties and inconveniences of irregular and unpredictable employment. Unfortunately, available data do not permit confident estimates of the proportion of the differential attributable to differing seasonal patterns.

The relatively stable employment experience of regular residential workers also suggests that some of the problems of formal training programs in union construction should not impinge as severely in homebuilding. There is, in the first place, less prospect of large numbers of unemployed trainees during the winter months, a prospect which leads many unions to limit the size of their apprenticeship classes. There is, furthermore, less of an inducement to trainees to drop out in favor of steadier income elsewhere. Finally, the lower incidence of seasonal layoffs serves to reduce the probability that the individual employer will lose a worker in whom he has invested training time and effort. The upshot here is that the management of formal industry-wide training programs in homebuilding would appear to be somewhat easier than in more volatile sectors of the industry. At the same time, however, as we shall see in the following chapter, there are other features of homebuilding—not the least important of which is the absence of the union input on training—which have rendered this aspect of labor market operations somewhat less responsive to the industry's needs.

CHAPTER IV

Manpower Training and Development

Seasonal fluctuations constitute only one element in the over-all labor market function of maintaining a labor supply to meet manpower needs. Even if seasonal movements were negligible, the market would still be required to respond to demand changes which occur over a longer period of time. These long-term patterns, moreover, generate somewhat different problems and responses from those associated with seasonality. In the first place, although the total demand for construction labor has risen steadily over time, the magnitude (and even the direction) of change is large and difficult to predict from year to year and from locality to locality. In addition, to the extent that additions to the labor force are expected to be more or less permanent, the kinds of temporary expedients described in the previous chapter are not directly germane. Instead, the industry must develop or attract sizable numbers of workers who are trained in construction work. In short, the market must plan with some degree of accuracy for a demand which is notoriously uncertain.

Long-term manpower needs are not, of course, entirely unpredictable. A substantial proportion of job openings are created by the natural attrition of the work force—deaths, retirements, and mobility out of the industry. Although net mobility tends to be a function of future economic conditions in other industries relative to construction—hence presenting serious forecasting problems—deaths and retirements can be estimated with some precision. In construction, it appears that such openings represent an appreciable portion of total needs. In 1965, for example, the Bureau of Labor Statistics projected that of 670,000 openings for *craftsmen* in the construction industry over the next decade, nearly two-thirds would be attributable to deaths and retirements, with the remainder created by net growth of the labor force.[87]

At the same time, much of the future need is exceedingly difficult to project accurately. In 1967, for example, the New York State Department of Labor projected construction employment in the state would rise from 254,900 in 1965 to 294,800 in 1970 and 329,000 by 1975,[88] but the actual figure for 1970 was 267,600.[89] For Buffalo, the projections were 19,200 to 22,700 and 25,300 with the actual 1970 level at 19,000. Any new openings implied in these figures, moreover, represent only net growth and do not take into account workers who migrate out of the industry. Anticipating mobility pattern, therefore, may well be even more hazardous than forecasting total demand. Even with the most sophisticated statistical techniques, the accuracy of the forecasts will still depend on a host of critical assumptions regarding the level and composition of construction activity and the health of the economy in general, among other factors.

The need for better demand forecasting and more use of available techniques have been explored in detail elsewhere.[90] Estimating not only total demand, but also such other variables as the occupational *composition* of demand, trends in output per man-hour, and the utilization of labor (e.g., turnover) is tricky business indeed. Furthermore, the forecasts must reflect differences in individual local labor markets, since relative stability in total demand may reflect growing shortages in one area along with growing unemployment in another. Clearly, however, the task of anticipating the need for bricklayers in

---

87 Allen F. Salt, "Estimated Need for Skilled Workers, 1965-75," *Monthly Labor Review,* Vol. LXXXIX (April 1966), p. 368. The forecast did not include any estimate of net mobility.

88 New York State Department of Labor, Division of Research and Statistics, *Manpower Directions: New York State, 1965-75,* Volume II (March 1968), p. 4.

89 U.S. Bureau of Labor Statistics, *Employment and Earnings: States and Areas, 1939-70,* Bulletin 1370-8 (Washington, D.C.: Government Printing Office, 1971), p. 377.

90 J. T. Dunlop and Daniel Quinn Mills, "Manpower in Construction: A Profile of the Industry and Projections to 1975," in *Report of the President's Committee on Urban Housing: Technical Studies* (Washington, D.C.: Government Printing Office, 1968), Vol. II, pp. 241-286; and Mills, *op. cit.*, pp. 89-118.

Buffalo is more difficult than that of estimating the future demand for construction workers in the United States. Yet for masonry contractors in western New York, the former is obviously a relevant question.

The burden of preparing for manpower requirements some years in the future bears more heavily on construction labor markets than on most others. Year-to-year changes in construction employment are typically greater than those in industry generally. On the local level, these changes may be great indeed. In the Albany, New York, area, for example, construction employment rose by over 75 percent between 1963 and 1970 (largely attributable to a massive state government project), but total employment there increased by only 19 percent.[91] In part, these prodigious expansions can be met by attracting workers from other areas, but in the event of a downturn the reallocation of the labor force is not entirely smooth. In Buffalo, construction suffered particularly heavy unemployment during much of the 1960's because of the buildup of the labor force in the 1950's to man the Niagara Power Project.

This problem alone would not be as severe in other industries because the need is not simply one of attracting bodies to fill job openings. Since one-half of the construction work force is classified as craftsmen,[92] there must be lead time to train substantial numbers of new workers. Although many have argued that most existing apprenticeship programs are too long, there is little doubt that a qualified mechanic cannot be created overnight. Yet if training programs base their operations on overly optimistic projections of conditions years hence, they will be producing skilled workers who cannot be readily absorbed. In short, not only is it difficult to forecast demand, but the need for accurate forecasts is particularly great in construction.

The bulk of this chapter is devoted to an examination and evaluation of the mechanics by which the supply of skilled labor in construction, and particularly in the homebuilding industry, is generated to meet the needs of attrition and growth. As we shall see, the industry relies principally on its own labor market processes and institutions, both formal and informal, to perform this function. In the union sector, formal training centers

---

[91] U.S. Bureau of Labor Statistics, *Employment and Earnings: States and Areas, op. cit.*, p. 387.

[92] In manufacturing, by contrast, only about 19 percent are craftsmen.

upon the apprenticeship system, although it has become clear that apprenticeship is only one of a number of avenues to skill acquisition. In residential construction, formal training appears to be appreciably less pervasive and is in fact virtually nonexistent in many labor market areas. It will be argued below that this situation has important implications for the development and retention of a skilled work force in homebuilding.

SKILL DEVELOPMENT IN THE HOMEBUILDING INDUSTRY

As noted in Chapter I, the best source of information on the background of skill acquisition is the worker himself. Since the research here is based entirely upon responses from employers and industry spokesmen, the findings must necessarily suffer from the respondents' incomplete knowledge of the workers' backgrounds. Nevertheless, the survey does permit certain conclusions about the training devices utilized by residential employers themselves, and even indirect accounts of workers' training elsewhere should provide some tentative conclusions or hypotheses for subsequent testing. The Erie County responses will be supplemented with brief accounts of the experience in other cities, and certain training efforts undertaken nationally under the sponsorship of the National Association of Home Builders will be discussed.

Experience in Erie County

In Erie County, there is no formal industry-wide program in the residential sector for the development of construction skills, a situation which is more the rule than the exception across the country. If workers receive any formal training at all, they receive it (1) outside the construction industry altogether; (2) in the union sector; or (3) from individual residential employers. The number of residential workers who have received formal training from any of these sources appears to be small. For the most part, skills are acquired informally and haphazardly through experience on the job. The result is that few workers develop a broad range of skills at their craft, and many can function only under the careful supervision of a foreman or the employer himself.

To some extent, the nonunion character of residential construction lends itself to the utilization of large numbers of partially trained workers. Without jurisdictional limitations and the requirement of uniform wages, the employer can divide many jobs into fairly simple, repetitive tasks and can hire marginally skilled workers to perform them at rates commensurate with their talents and productivity. Under these circumstances, however, the role of the supervisor is critical, both to assume overall direction and planning of the job and to handle those tasks outside the routine. When jobs are plentiful in the higher-paying union sector, however, the builder stands in some danger of losing his best workers with obvious consequences to the efficiency and cost of his operations.

That the bulk of the work force is not trained in any systematic way is demonstrated by the responses of Erie County employers to various questions on the survey. These questions fall into three groups: (1) those relating to the training efforts of the builders themselves; (2) those relating to the training backgrounds of their workers; and (3) those relating to employer attitudes toward the prospect of a centrally run formal training program in Erie County.

Training by Employers. Although most employees of residential builders and subcontractors have some skill when they are hired, an appreciable number are hired with no prior training. In Erie County, 28 percent of the respondents indicated that over one-half of their new hires had not had any training, and only 23 percent reported that their new workers invariably possess some degree of skill (see Table 31). For the most part, untrained workers were taken on chiefly because skilled workers were not readily available, although some employers did express a preference for the neophyte. For example, one cement contractor said, "On occasion we have hired from other companies like ours, but these workers never last. They don't do things our way or are no good, which is why we prefer to get them green as grass." Furthermore, the propensity to hire untrained workers varies with the skill requirements of each job. In the relatively demanding fields of plumbing, heating, and electrical work, 17 of 20 respondents (85 percent) said that one-half or less of their new hires were unskilled; the corresponding number for the other subcontractors was 19 of 34, or 56 percent.

Most new workers are expected eventually to perform work requiring some degree of skill. Employers were asked the following question: "Of your new workers who are hired without prior training, what percentage are given skill training on-the-

TABLE 31. *Homebuilding Industry*
Percent of Workers Hired Without Prior Training, By Type of Employer
Erie County, New York, 1971

Percent Hired Untrained	Total	Type of Employer	
		Builder	Subcontractor
0	20	8	12
1-25	24	11	13
26-50	18	7	11
51-75	6	1	5
75-99	6	1	5
100	12	4	8
Total	86	32	54

Source: Erie County survey, 1971.

job and what percentage are used *only for work requiring no particular skill?*" Of the 71 employers who responded to this question, 52, or 73 percent, indicated that none of their new hires were intended for unskilled work exclusively. Generally, subcontractors were somewhat more likely than builders (and the basic trades subcontractors more than the mechanical trades subcontractors) to provide some type of training. Of the 19 respondents who reported any hiring of workers for unskilled work alone, 15 were either builders or cement contractors.

Apart from training on the job, few employers undertake to provide their workers with any kind of systematic instruction. Only five respondents reported a practice of giving their new workers job-related instruction outside the workplace either by themselves or by a skilled employee, and only one said that he provided workers with books or manuals to study. Eight others indicated that they paid the tuition expenses for employees attending evening classes in a job-related subject. In all, 55 of 70 respondents were not engaged in training efforts other than on-the-job.

The term "on-the-job training", moreover, is imprecise. By one standard, it is a systematic process including progressive exposure to various elements of the trade, repeated performance of tasks under the watchful supervision of a qualified instructor, and periodic evaluation of the trainee's accomplishments. Such a process is not only costly, but it also requires competent instructors to supervise the work. Since residential contractors typically employ relatively few broadly skilled craftsmen, who themselves must turn out finished work and therefore have little time to devote to overseeing others, it seems likely that in this context on-the-job training often refers merely to the trainee's randomly observing the work of others. Although it cannot be denied that such observation may well ultimately lead to an ability to perform some skilled work, it does not necessarily follow that the process constitutes "training." Training presumably involves a "trainer," and unless the journeyman actually takes time from his own work to instruct and guide the learner in some logical sequence of tasks, there is, strictly speaking, no training. Another criterion of training is the inculcation of a capacity to perform without close supervision. Here again, the tasks learned randomly are likely to be highly specialized, so that careful supervision is still needed to coordinate the various narrow functions that make up the job. It is important, in short, to note the distinction between purposeful training and haphazard learning, for it is the former which will turn out the qualified journeyman most efficiently.

It was apparent that the individual employer frequently lacks the resources to provide broad, systematic instruction to his untrained or marginally trained workers. If the trainee is to learn, he cannot be used exclusively for simple repetitive jobs in which his productivity is commensurate with his wage. He must spend time during which he is not producing much of anything or during which his output is low and perhaps replete with costly mistakes. A skilled and highly paid worker must also interrupt his own work to supervise and instruct the trainee. Ideally the trainee should be exposed to the principles of construction methods, the materials used, and some sense of the progressive erection of a building—instruction which is best furnished off the job. All these factors represent cost, and even if the employer were willing to assume it, he has no assurance that the newly trained worker will remain long enough for the employer to benefit from his training. Clearly, then, the relative absence of systematic training in residential construction is a function less of the parsimony of individual employers than of the failure of the industry as a whole to develop locally an organizational structure and commitment to skill generation.

Training Background of Workers. The validity of the responses here is open to question, particularly in light of the practice of most residential employers to base their hiring decisions on tryouts rather than on rigid minimum standards of education or experience. Nevertheless, it is not unreasonable to expect employers to be familiar with the backgrounds of their workers, and so their general impressions as to the kind of training their workers have received are of some value. This appears to be so in light of the responses to previous questions.

In the Erie County survey the training background of workers was elicited by the following query:

There are a number of ways your workers (past and present) might have obtained training before you hired them. Would you please give us a rough estimate of the percentage of your workers who have utilized each of the sources of training listed on this card.

a. Picked it up on the job from other homebuilders or subcontractors.
b. Picked it up on the job in other branches of construction.
c. Picked it up on the job in a different industry (e.g., farming, manufacturing, etc.).
d. Went through a union apprenticeship program.
e. Took courses in a trade school or technical institute.
f. Learned from friends or relatives.
g. Learned in the military.
h. Others (Please specify).

There were 77 respondents to this question. (Several employers did not respond because they had indicated previously that all their workers were hired untrained); their replies are summarized in Table 32. The most striking distinction in this table is between "picked it up on the job from other homebuilders or subcontractors" and all the other categories. It is apparent that formal training of any kind is extremely rare in residential construction in Erie County. The survey also revealed that most residential employers appear to rely heavily on the informal instruction provided by other employers, although few knowingly derive any substantial benefit from training sources external to the industry.

At the same time, it is obvious that a young worker has to begin somewhere, and *all* new hires in the industry could not have picked up their skills from "another" employer. It is appropriate, therefore, to examine the alternative sources of training in somewhat greater detail.

Other Sectors of Construction. A small number of residential workers learn their trades in other branches of the industry,

TABLE 32. *Homebuilding Industry*
Training Background of Workers as Reported by Employers
Erie County, New York, 1971

Source of Training	Total	Number of Respondents				
		Percent of Workers Receiving Training				
		0	1-25	26-50	51-99	100
Other residential employers (on-the-job)	77	7	7	23	13	27
Nonresidential construction (on-the-job)	77	48	18	7	2	2
Other industries (on-the-job)	77	60	11	3	3	—
Apprenticeship	77	65	7	4	1	—
Trade school or technical institute	77	51	21	4	1	—
Friends or relatives	77	59	11	4	2	1
Military	77	70	5	—	1	1

Source: Erie County survey, 1971.

either informally on the job or through an apprenticeship program (see Table 32). As noted earlier, the builders in Erie County at one time operated under an agreement with the Carpenters Union, and six of the twelve respondents who reported that some of their workers had been union apprentices were mainly employers of carpenters (either builders or subcontractors). The remaining six contractors were in heating (3), painting, cement work, and plumbing. Carpentry is also the chief craft among those who reportedly picked up skills on the job in nonresidential construction: employers of carpenters constituted 14 of 29 such respondents. Others were cement (4), plumbing (3), heating (3), excavating (3), and painting (2).

Other Industries. In the 1968 Syracuse study, it was observed that "although interindustry mobility provides construction with a substantial number of workers, the extent to which these workers are actually trained in other industries is yet another question."[93] The findings here tend to confirm that view. Only

[93] Howard G. Foster, "Nonapprentice Sources of Training," *op. cit.*, p. 22.

17 respondents noted other industries as a training ground for any of their workers. By far the most commonly mentioned "other" industry was farming, although a few employers alluded to maintenance work in a manufacturing plant. The major crafts employed by the 17 employers in this group were carpenters (7), plumbers (4), machine operators (2), heating men (2), cement workers (1), and electricians (1).

Trade or Technical School. A total of 26 respondents reported that some of their workers had received job-related instruction in a classroom setting.[94] Although this source of training was the most common avenue outside the industry itself, it is appropriate to recall an observation from the previous study mentioned above:

> The point here is by no means that in-school vocational education is a useless form of training, but rather that in current practice it often does not complete the worker's training, that it usually must later be combined with other avenues of skill acquisition. . . . Thus vocational education . . . seems to have been only the first step of several in the attainment of skilled craftsman status.[95]

Some respondents offered corroboration of this conclusion. One excavating contractor noted that his workers "didn't get enough training to run small machines. Not qualified enough." A heating contractor observed that "we had a boy once who was taking sheet metal in high school, but it didn't have anything to do with what we were doing." The most common trades mentioned in this connection by employers were carpentry (12), heating (4), and plumbing (3).[96]

Friends and Relatives. It is important to note that in an industry which is often characterized as "nepotistic," 59 of 77 employers (77 percent) believed that *none* of their workers had acquired their trade in this manner. Of the remaining 18 respondents, 7 were builders or carpentry subcontractors and 6 were cement contractors.

Military. On this point Table 32 shows that of the five employers specifying the military as a training ground for some

---

[94] In addition to trade school and technical institutes, these included service schools run by materials suppliers.

[95] Foster, "Nonapprentice Source of Training," *op. cit.*, p. 25.

[96] For a discussion of the need for greater cooperation between the industry and vocational schools, see Mills, *op, cit.*, pp. 202-206.

of their workers, four were in the related fields of plumbing and heating work.

The survey of Erie County builders and subcontractors illustrated that the *primary* means of skill acquisition, both formal and informal, is training within the construction industry itself. The principal difference is that in residential construction there is very little formal training, so that most training is through the informal and unstructured acquisition of skills flowing from the production process itself. We have already noted the possible unreliability of this method of skill development; we turn now to employers' views on the desirability of more formal training in homebuilding.

Employer Attitudes Toward Formal Training

For the most part, residential employers in Erie County agreed to the desirability of a structured, centrally run training program for nonunion construction workers. This attitude was related primarily to a feeling on their part that qualified workers are in short supply. Asked whether they felt "that there is or is not currently a shortage of qualified workers," 31 of 33 builders (94 percent) and 45 of 56 subcontractors (80 percent) replying responded in the affirmative. Respondents were then asked the following question:

> Do you feel that it would or would not be a good idea to establish an apprenticeship program of some sort (perhaps administered by the Builders Association) which *combines* on-the-job training with classroom instruction?

To supplement these responses, we attempted to gauge the strength of the employers' feelings by asking whether they would be willing to contribute to the operation of an apprenticeship program. Some erosion of the support for such a program was detectable when expressed in terms of cost. Forty respondents were willing to make a contribution and eight more were unsure. There was no clear difference between larger and smaller employers with respect to their views on formal training.

By any of the foregoing measures, however, builders were generally more favorable toward formal training than were subcontractors. A part of this difference is undoubtedly attributable to a distrust of the Builders Association on the part of subcontractors, most of whom are not Association members. Primarily, however, those with negative attitudes appeared either

TABLE 33. *Homebuilding Industry*
Attitudes Toward the Desirability of an Apprenticeship Program In Residential Construction, by Type of Employer
Erie County, New York, 1971

Attitudes	Total	Type of Employer	
		Builder	Subcontractor
Desirability:			
Yes	62	29	33
No	15	2	13
No shortage of workers perceived	9	1	8
No answer	4	1	3
Total	90	33	57

Source: Erie County survey, 1971.

to deny the necessity for a central program with related instruction, or more commonly, to fear the loss of trained workers to other builders or commercial contractors. The following remarks illustrate some employers' negative attitudes toward apprenticeship training.

—Builders and subcontractors would spend the money to train someone who would leave if he didn't like the work. Go into a plant. Would be a waste of money.

—I don't feel I have the time to train on the job and there is no guarantee he will stay with you after he gets the training.

—Most men would not be interested in classroom type instruction. Can learn better actually on-the-job.

—Might end up training people for commercial work.

—Certain general contractors would get the new trainees and the rest would get left out. Also maybe then the new trainee would leave to join the union.

—What would I get? What would happen if these fellows who get trained would stay a while and as soon as they know what it's all about they would go off for themselves and become your competitor.

—I think we have had boys work with us and they get a lot more out of it. I don't think any classroom would help.

—I don't think the Builders Association could hold them. They'd go to commercial once they learned the skill. Homebuilding can't pay the wage that commercial does.

These comments reveal that some employers do not understand the educational principles underlying apprenticeship, betraying the erroneous notion that classroom instruction is its only component. Nevertheless, it remains true that nearly one-half of the respondents were willing to subsidize a formal training program, and three-fourths of them approved of the idea in principle. Before we draw any conclusions concerning the feasibility of an apprenticeship system in nonunion construction, we must examine experience with formal training arrangements in other areas.

TRAINING IN RESIDENTIAL CONSTRUCTION OUTSIDE ERIE COUNTY

Interviews with industry spokesmen in other communities revealed that the absence of formal training programs in Erie County is not atypical. In some of these areas, limited training programs were undertaken on a sporadic basis with varying degrees of success. Nowhere, however, were such programs sustained over a period of time and the number of trainees involved was invariably small.

Rochester

The Executive Officer (E.O.) of the local Builders Association reported that the majority of young workers start out as laborers or helpers on summer and part-time jobs and "work their way up." There is no formal relationship with any vocational school because presently there is no way to guarantee jobs for graduates. The E.O. did not feel that military service constituted a significant training ground. "Quite a few" workers enter the industry with a farm background, although only about 5 to 10 percent of all new entrants start with *any* prior training in construction work. There has, however, been some recent influx of workers from the commercial sector as commercial activity slows.

The E.O. also described an experimental program run by the Association four years earlier. Ten masons and 25 carpenters were given intensive manipulative training by three skilled masons and ten skilled carpenters. The trainees received $1.65 an hour plus their tools, and in two weeks they built a wall and two houses. The Association had guaranteed jobs for the trainees, but the builders did not follow through. Placement was very difficult, and none of the trainees have remained in the industry. The program, not surprisingly, has been terminated.

A large builder in Rochester claimed that union workers occasionally set themselves up as residential subcontractors, particularly during a strike (presumably on a temporary basis, although the builder did not specify). A few subcontractors, moreover, received their training from the Kodak Corporation which has a large construction operation of its own in Rochester.

Syracuse

In this city, informal progression was again emphasized: "Kids start from scratch and work up." Some builders, according to the E.O., "have their own programs"; however, one builder has worked through the Board of Cooperative Educational Services (BOCES),[97] donating a building lot and materials for students. In the E.O.'s opinion, vocational schools or training in other industries are not significant factors. This same picture was also drawn by individual builders interviewed in Syracuse.

One builder described the BOCES program sketchily. It included some classwork, but involved mostly on-the-job training during the summer months. It was offered for three years, during which time this builder was the only one to provide material assistance. There was only one experienced supervisor who was paid by BOCES. The builder did not know if the program would be continued or whether those completing classes had been successful in finding jobs.

Boston

In Boston also the stress was on "picking up" skills. The Builders Association plays no role in the process, although a few individual builders have arrangements with local trade and technical schools. Workers in the mechanical trades are more likely to have started in a post-high school technical institute and then found jobs as helpers. On the whole, the Boston E.O. felt that training has not presented a problem to the homebuilding industry.

---

[97] Board of Cooperative Educational Services (BOCES), a country-wide consortium providing specialized kinds of instruction—music, shop, etc.—for area school districts.

Providence

The Providence area boasted the most ambitious formal program encountered, although to put it in proper perspective very few workers have undergone formal training of any kind. The Rhode Island Builders Association (RIBA) has sponsored the building of nine houses in the past three years by students at three area vocational schools. Students attend classes in the morning and work in the afternoon. The RIBA provides the site, the materials, and the supervision through member builders. Crews consisting of 12 to 15 students and one supervisor perform all building functions except excavation, bricklaying, plumbing, and electrical work. The houses are then sold to provide funds for further construction. One project for 1972 was scheduled a year in advance.

A total of 100 to 120 students have graduated from this program. It is not known, however, how many remained in the industry, although some of those who left had been drafted.

Cleveland

The only formal training by homebuilders in Cleveland is through the union apprenticeship programs. Spokesmen for both the Homebuilders Association and the commercial contractors' association agreed that homebuilders seldom use apprentices even though they contribute to the programs' funds. One factor affecting this is the recent slowdown of construction which has resulted in cutbacks in most of the apprenticeship programs. Homebuilders, according to the HBA spokesman, cannot afford to employ apprentices at the union rate because trainees do not produce as efficiently as the other workers. By and large, training is of the informal "pickup" variety, although there is more in-mobility from the union sector since there is no wage differential. Of course, some proportion of those starting out in commercial construction have been apprentices.

Other Areas

The picture obtained by interviewers from the Industrial Research Unit in Baltimore, Denver, Indianapolis, and Philadelphia was not substantially different from that depicted in the cities of Rochester, Syracuse, Boston, Providence, and Cleveland. The only formal industry-run program was found in Denver, part of the federally funded Carpentry Apprenticeship Program. There was no report of formal training efforts by individual contractors. Most respondents noted that young workers often began in vocational schools, although there was no mention anywhere of work-study arrangements between employers and schools. The training was invariably characterized as "on-the-job," and its nature probably best described by one suburban Baltimore builder: "On-the-job training, but not in the formal sense. Knowledge and skill are picked up by observation." On the question of initiating and supporting formal programs through the local builders association, the reaction was mixed with most affirmative responses hedged. On the whole, these interviews tended to corroborate the findings in Erie County and elsewhere, and they indicate that formal training in homebuilding is virtually nonexistent except when organized under the auspices and with the help of the National Association of Home Builders.

TRAINING NATIONALLY

There is very little information on the extent of formal training in homebuilding nationally. The experience of the few areas discussed in the preceding section suggests that formal training is much more the exception than the rule. The 1969 survey of the National Association of Home Builders membership lends support to this conclusion. Only 11 percent of the respondents answered the following question affirmatively: "Do you participate in any local manpower training programs?" As might be expected, the percentages were somewhat higher for large volume and multifamily builders than for single-family builders producing relatively few units.[98] There is no indication of how many of the manpower training programs cited took the form of union apprenticeship.

Carpentry Apprenticeship Programs

In nonunion homebuilding, the major source of formal training appears to be carpentry apprenticeship programs sponsored by the NAHB through several of its local affiliates.[99] Since

---

[98] Sumichrast and Frankel, *op. cit.*, p. 202.

[99] Much of the material here is based on interviews with NAHB staff members and a brief discussion of the Association's training efforts in *NAHB Journal of Homebuilding*, Vol. XXVI (January 1972), pp. 81-85.

1967, the Departments of Labor and Health, Education and Welfare have given financial support to the Association for the development of preapprenticeship programs lasting a total of 22 weeks. These programs were intended to encourage the establishment of regular four-year apprenticeships similar to those in the union sector but emphasized skills with particular application to homebuilding. By 1973, four-year programs had been initiated in about forty small- and medium-sized cities (some of which were actually being administered on a state-wide basis), including eighteen in which preapprenticeship programs were also in operation. Through mid-1972 the government had granted $2.2 million for these efforts which included instruction not only for trainees but also for journeymen who were to supervise the trainees on the job.

These individual programs are established through subcontracts with the local HBA, and are administered by committees consisting of representatives from the builders, participating vocational schools, and federal or state apprenticeship agencies. When a local program is funded, it hires a full-time coordinator and provides for both classroom and on-the-job training. Most of the money is used for the coordinator, classroom instructors, and textbooks. The coordinator himself recruits the trainees—15 to 20 in each class—usually through a local community agency like the Urban League or the State Employment Service. As noted, the program itself is conceived as a pilot project, en route to the ultimate development of builder-run four-year apprenticeship programs which can be registered with the U.S. Bureau of Apprenticeship and Training (BAT) or the equivalent state agency. In fact, by 1973 almost all the four-year programs had been so registered.

The classroom phase, lasting six weeks, covers 240 hours of instruction in the following subjects: mathematics for carpenters (24 hours); lumber (16 hours); other construction materials (16 hours); blueprint reading (24 hours); carpenter's tools (24 hours); safety (8 hours); laying out for construction (24 hours); foundations (16 hours); and framing (88 hours). Trainees receive no stipend for this part of the program, although they are eligible for training allowances from other government sources. The training itself is usually conducted in local vocational or trade schools, and in some cases the cost of instruction (more precisely, the instructor's stipend) is assumed by state or local agencies.

The on-the-job training phase lasts 16 weeks, during which trainees are placed with a particular builder. The actual training and the trainee's wages are the responsibility of the builder himself, although the program coordinator is expected to keep in touch with both employers and trainees and to ensure that instruction in the various facets of carpentry is actually being given. The experience of the NAHB has been that most trainees complete the program; of the 2,331 trainees who started in the first four years of the program, 1,850 were still in the trade in early 1972.[100]

The preapprenticeship program is intended in large measure to attract minorities to the homebuilding industry. According to the NAHB, of the 3,500 persons who entered programs through early 1973, some 70 percent were classified as either minority or disadvantaged. There is no indication, however, whether the retention rate of minorities is greater or less than that of other trainees. In the aggregate, about 70 percent of those beginning preapprenticeship go on to a four-year program, but the proportion who actually complete the full apprenticeship program is not known.

The journeyman upgrading program for carpenters is generally conducted in conjunction with an apprenticeship program, and by the end of 1972 it had reached 165 journeymen. This program consists entirely of classroom training with classes held one evening each week for 14 weeks. The topics covered include residential blueprint reading, job cost estimating, leadership and supervision, and new technology.

Bricklayers Apprenticeship Programs

In addition to the carpentry program, the NAHB has instituted a bricklayers training program for military servicemen who are about to be released from active duty. This is essentially a six-week preapprenticeship program conducted at various military bases throughout the country, and is funded by grants of some $700,000 from both the Departments of Labor and Health, Education and Welfare. During its first two years (1970-1972), 3,100 trainees completed the course.

The program's curriculum consists of 240 hours of both classroom instruction and manipulative training. Class subjects include methods, use of tools, mathematics, and blueprint read-

---

100 *Ibid.*, p. 82.

ing. Trainees also receive practice at laying of both block and brick, usually by a group's undertaking some small project on the military base.

The principal problem inherent in this program is the placing of trainees. Unlike the carpentry programs, there is no commitment that an opening exists once a trainee has finished. The NAHB attempts to find jobs through its local affiliates in the city to which the trainee intends to return. The NAHB estimates that only one-half of the trainees are actually placed on jobs or in regular apprenticeship programs, although this percentage may be an understatement since the Association is not always notified when a trainee is placed. There is also no indication, however, of how long even those trainees who are placed remain with their employers or in the trade.

Presently, not enough information is available on these NAHB programs to permit an in-depth evaluation of their performance. Still, a few general observations may be made. It does not appear that these efforts, however well-conceived and executed, operate on a very large scale. If the dropout rate from union programs is any guide, then the number of apprentices completing the four-year homebuilding programs in carpentry has apparently averaged on the order of 300 per year for the past six years throughout the country. The 40 areas presently participating in such programs constitute only a fraction of the labor market areas in which residential construction is predominantly nonunion. The masonry program is numerically larger, but its limited duration and lack of guaranteed placement raises doubts about the extent to which it has contributed to the development of fully qualified bricklayers. Finally, carpenters and bricklayers together constitute only a little more than one-half of the skilled work force necessary to build a single-family home; there appear to be no extensive formal training programs for electricians, plumbers, operating engineers, or other craftsmen who work in the residential sector.[101]

[101] It should be noted, however, that the Associated Builders and Contractors, a predominantly nonunion group comprised of both residential and nonresidential firms, has initiated training programs for some of these trades. These programs will be examined more extensively in the forthcoming study of open shop construction.

APPRENTICESHIP IN THE BUILDING TRADES

Most *formal* training of construction workers occurs through union-management apprenticeship programs which combine on-the-job training with (typically) 144 hours of classroom instruction. Most programs in the building trades conform to standards set down by the U.S. Bureau of Apprenticeship and Training and include: the learning of a series of manipulative skills according to a set schedule; in-school related instruction designed to supplement on-the-job training with the theoretical foundations of the trade; a progressively increasing wage schedule, usually expressed as a percentage of the journeyman wage; on-the-job supervision by qualified journeymen; periodic evaluation of the apprentice's performance; and an agent whose task it is to administer the program and ensure that each apprentice receives the requisite instruction both on and off the job. Upon completing his program, in most trades after three or four years, the apprentice is expected to have received a well-rounded exposure to all facets of the trade under study.

Construction programs are administered by local Joint Apprenticeship Committees (JAC's) which are composed of an equal number of representatives (usually three or four) from labor and management. The JAC as a whole is responsible for the routine functions of the program such as interviewing prospective apprentices and allocating them among the various employers. It seems safe to conclude, however, that the key decisions as to the administration of these programs are effectively in the hands of the union. Programs are funded by a small assessment on each employer for every man-hour of labor he employs, regardless of whether he himself uses apprentices. These funds are used for activities such as advertising openings, hiring instructors, providing tools and materials, and, in some cases, paying apprentices a stipend while in school. For on-the-job training, the apprentice, in the ideal situation, shifts from one employer to another, ensuring him exposure to all aspects of the craft.

The rules and regulations governing apprenticeship are formally determined by the local JAC. All the crafts, however, have JAC's at the national level, and these periodically promulgate "suggested" standards for the administration of local programs. These standards cover topics such as age limitations, length of the program, probationary periods, wage schedules, and methods of evaluating apprentice performance. Although one may occasionally encounter local deviations from these standards, they are, by and large, followed. The national standards, however, do not address themselves to questions such as the number of apprentices in the program, the maximum number an employer may hire, or the ratio of apprentices to journeymen.

It is difficult to assess the extent to which apprenticeship contributes to the overall acquisition of construction skills because comprehensive data on both apprenticeship and employment patterns are lacking. It seems clear, nevertheless, that only a fraction of construction workers in the aggregate receive formal training through apprenticeship. For example, the Bureau of Labor Statistics study cited earlier estimated that in the decade 1965 to 1975 about 41,000 openings per year would result from deaths and retirements alone.[102] In recent years, apprenticeship programs for construction trades have graduated about 20,000 persons annually.[103] The figure on openings, moreover, takes no account whatsoever of net growth or out-mobility. Of course, the ratio of openings to graduated apprentices varies greatly from craft to craft, but it is certainly safe to say that no craft fills all its openings with apprentice-trained journeymen.

At the same time, apprenticeship is not designed to supply the industry with a full complement of rank-and-file journeymen. Not all craftsmen need the extensive training provided by apprenticeship. Every craft includes tasks which can be learned very quickly. The carpenter who builds forms for concrete does not need to be as skilled as the one who installs wood trim. Many of these functions can be, and indeed are, picked up by means other than apprenticeship. Thus, although there is certainly a need for workers with a broad range of skills to assume general supervision of the work and to handle the more difficult problems and tasks, it is naïve to presume that *all* workers must have this kind of well-rounded training. Indeed, apprenticeship has traditionally supplied a disproportionate number of foremen, superintendents, and even contractors to the industry, as well as the key journeymen who form the basic construction work force and who typically receive more hours

[102] Salt, *loc. cit.*

[103] U.S. Department of Labor, *Manpower Report of the President* (Washington, D.C.: Government Printing Office, 1972), p. 271. It should also be noted that this number includes some workers outside the construction *industry,* although the figure on openings does not.

of employment during the year than the average craftsman.[104] Unfortunately, there is presently no way to fix accurately the size of this "basic" work force.

The employer's role in the apprenticeship process frequently receives less attention than that of the union, but it is far from unimportant. To the individual contractor, the hiring of an apprentice may quite reasonably appear to be a bad investment. At the beginning at least, the apprentice is likely to be paid more than his productivity warrants. In addition to the apprentice's wage, the contractor must pay for a journeyman to supervise the work and for any wasted materials resulting from the apprentice's inexperience. Furthermore, he is expected to keep shifting the trainee from one aspect of the job to another, thus failing to capitalize on the experience gained by the apprentice at any one task. It seems to be generally acknowledged that the employer does not begin to recoup his investment until the second or third year, when the apprentice has been sufficiently trained to perform some journeyman's work at apprentice wages. Finally, the employer has no guarantee of holding the apprentice after training is completed; although the training may benefit the industry as a whole, he alone bears most of the cost. As a result of these considerations, employers often prefer to hire workers trained elsewhere than to train their own.

The performance of apprenticeship programs in responding to the changing manpower needs of the construction industry is limited by the unavailability or nonuse of reliable indicators of future demands, as well as by the lack of correspondence between employer interests and industry interests. Attempts to find a statistical relationship between apprentice numbers in the several crafts and various national economic indicators have not been fruitful.[105] Without the development and dissemination of sophisticated planning guides, long-term labor force adjustments will probably continue to rely heavily on other avenues of skill acquisition.

At the same time, the very existence of an established, ongoing formal training mechanism should not be minimized. It strains credulity to assume that a sustained period of rapid

---

[104] For the latest evidence on this point, see William S. Franklin, "An Analysis of Traditional Routes of Entry into Selected Construction Unions," Ph.D. dissertation, University of Texas at Austin, 1972.

[105] See especially Mills, *op. cit.*, pp. 223-230.

growth in demand would not eventually call forth a response from apprenticeship administrators. The mechanism itself, moreover, can be encouraged through better information systems and a more active role taken by national leaders in the industry to respond more expeditiously to changing conditions. Despite its limitations, the current apprenticeship system, by common consent of both employers and union officials, produces a competent, well-rounded craftsman, and comparisons with the ideal must be tempered by comparisons with the situation in which formal training is largely absent altogether.

Skill Development Outside Apprenticeship

Training in construction, other than the completion of an apprenticeship program,[106] is gained primarily through the informal unstructured acquisition of skills flowing from the production process itself. Such training is actually aided by the seasonal fluctuations in activity described earlier, for the employer is obliged at times to utilize unskilled and semiskilled workers in jobs at which they are only slightly experienced, and the practice thus gained is a vital part of learning. In the union sector, of course, craft jurisdictions impede the free flow of manpower from one occupation to another. Nevertheless, the avenues of occupational change remain partially open and the astute or fortunate worker may be able to take advantage of them. As Strauss has described it, "In informal, even underhanded fashion, men learn (or 'steal') a craft bit by bit without going through any period of formal training." [107]

Nonapprenticeship training for construction work may take place both within and outside of the industry.[108] Within the industry, the process generally operates through a gradual upgrading of marginally skilled workers, as they improve their abilities by observation and occasional direct experience. This upgrading may involve movement from an unskilled occupation to a skilled craft or movement within a craft. Intracraft upgrading is made possible by the inclusion in most trades of relatively undemanding jobs, so that the worker can start out

---

[106] Strauss, *op. cit.*, p. 327.

[107] *Ibid.*

[108] Foster, "Nonapprentice Sources of Training," *op. cit.*, pp. 21-26.

performing simple tasks and then progressively broaden his skills. This process also applies to the laborer who over time picks up enough knowledge so that when skilled workers are in short supply he is pressed into service by a contractor.

An understanding of the operation of the *internal* labor market in construction is vital to an understanding of how the industry responds to changing needs. These internal market mechanisms, however, do not appear to be radically different in the nonunion sector, although they do appear to operate more smoothly in the absence of craft jurisdictional restrictions. There is still a *key* distinction in the union sector. Informal training functions as a *complement* to apprenticeship, and it is generally not relied upon entirely to produce the broadly skilled journeyman or foreman around whom the work force is built. In homebuilding, however, the virtual absence of formal training in many labor market areas means that informal mechanisms must produce the cadre of highly skilled craftsmen who not only perform the most demanding work but also serve to instruct and upgrade other workers.

CONCLUDING REMARKS

An evaluation of training in homebuilding based upon informal absorption of skill and knowledge on the job must question whether it is adequate to supply the industry with the requisite number of broadly skilled workers. The answer to this question lies with the state of the market at any given point in time. When construction activity is low, the residential sector can function with a large number of partially trained, semiskilled workers under the direction of a few highly competent and knowledgeable craftsmen. These "key workers" can be developed by the industry itself, starting out as helpers and laborers and working up through their own ambitiousness, drive, and intellectual gifts and supplemented by a few apprenticeship-trained craftsmen from the nonresidential sector who for one reason or another choose to work in homebuilding.

When the pace of activity quickens, however, builders are likely to be the first to feel the pinch. As jobs open up in the union sector, jobs which may pay much higher wages and may even offer overtime as well as attractive fringe benefits, the nonunion builder's key people—those who can best command union wages—will begin to leave. Since the homebuilder depends heavily on the skills of his relatively small number of true craftsmen, his problem is especially severe. Not only is it necessary for him to train new workers, but the individuals upon whom the responsibility for training and supervising has rested may now be the very ones moving away. The difficulty, moreover, may be compounded if the abundance of jobs in the nonresidential sector is matched by an upsurge in the demand for homes. Under these circumstances, the need for skilled workers in homebuilding would be growing although trained craftsmen are not being released in other sectors.[109]

The particular sensitivity of homebuilding to changes in demand for various kinds of construction products is, perhaps, the major justification for a more structured system of skills development, one which can respond to such changes as they occur. The residential sector should not have to depend on training programs elsewhere. Unions and their employers are not apt to expand their apprenticeship programs to supply the nonunion sector with trained workers. Builders will benefit from external training sources only when those decision makers miscalculate their need or when already trained workers there are released because of slackening demand. The uncertainty of this source of supply is reflected in the generally favorable attitude of the Erie County respondents to the principle of a formal training program of their own.

The development of formal training programs in homebuilding must come primarily at the initiative of the industry itself, and the attitudes of the Erie County respondents, as well as those in some of the other areas visited, suggest that the will is there. At the same time, local associations do not yet have the resources or the expertise to undertake such a project. As we have seen, there are external costs to training, thus the financing of any worthwhile training program needs to be industry-wide. To generate the necessary funds, HBA's will have to raise money through special assessments of their membership. The assessments should be scaled according to the size of each paying member, so that the cost bears some relationship to the potential benefits derived.

Any training program, and particularly one in its seminal stages, has need for a full-time coordinator and administrator.

---

[109] As of this writing (1974), of course, homebuilding is depressed. When the industry recovers, however, the issue discussed above will be far from academic.

There must be an individual who is responsible for recruiting trainees, placing them with employers, monitoring their progress, and arranging for off-site related instruction. Since resources are limited, there should be attempts to enlist the cooperation of local vocational education officials to provide the relevant supplementary classroom experience. Such an arrangement would also assist the vocational education establishment to ensure that their curricula are consistent with the needs of employers. The carpentry program of the NAHB is a significant move in these directions, and it merits continued support. At the same time, ways should be sought to develop training efforts for other crafts through other employer groups if necessary.

At the same time, there is some justification for training subsidies from federal manpower programs, since persons trained in homebuilding may well migrate to other industries. The NAHB preapprenticeship program is federally funded in its entirety. It is felt that some type of cost-sharing arrangement is preferable because it allows a given commitment to be spread further and also encourages the industry to devote its own resources to an activity which it recognizes as important. Manpower Development and Training Act monies could be used, for example, to pay part of the coordinator's salary or to supply books and materials to vocational schools for related instruction. Whatever the precise approach taken, the government should avoid creating the impression that training is free, but rather should use its resources to encourage the industry to develop its own permanent training institutions.

However established and administered, the industry would benefit from an ongoing mechanism through which it can respond to changing long-term manpower needs. Individual employers cannot be expected to shoulder the burden; it must be done on an industry-wide basis. None of this is to deny the difficulty of projecting such needs over time. We have already seen that union apprenticeship programs respond imperfectly, at best, to demand movements. At the same time, apprenticeship has assured a reasonably steady flow of well-trained journeymen into the work force, without whom the various types of informal adjustments described earlier would be less effective. Furthermore, the apprenticeship system at least provides a mechanism through which new demands can be met; if it does not work perfectly, it is, nonetheless, a base upon which improvements can be made.

The foregoing does not suggest that in the absence of more formal training the industry will be unable to function. Rather, the point is that, given the abrupt changes experienced over relatively short periods of time (housing starts increased by some 62 percent between 1970 and 1972)[110] and the danger of losing key workers to other sectors of the industry, the home-building labor market would be able to respond more expeditiously if it had a steady source of highly trained workers around whom the aggregate labor force expansion could be built. Although informal and unstructured training methods are necessary in an unstable market, they are most effective as a supplement to an underlying system of training which produces a base of well-rounded journeymen.

---

[110] U.S. Department of Commerce, *Construction Review*, Vol. XIX, no. 10 (October 1973), p. 14.

CHAPTER V

Matching Supply and Demand: The Hiring Process

The efficient operation of a labor market involves not only procedures for ensuring an adequate supply of manpower, but also a mechanism by which vacant jobs can be filled with the available workers. As with training, the job matching function can be systematic and formal or haphazard and informal. The principal criteria for evaluating the efficiency of any particular method or process should be the speed and accuracy with which employment information is disseminated and the freedom of both workers and employers to utilize that information. Thus one manifestation of an imperfectly functioning market is the simultaneous existence of job openings and unemployed workers. Although a certain amount of labor market friction is inevitable and measurement problems are severe, an empirical examination of the hiring process in homebuilding may allow some inferences as to the efficiency of manpower allocation in the industry.

The problem posed here, then, is one of meeting the demand for labor from an industry-wide or market-wide standpoint, but rather one of linking the individual job-seeking worker with a specific job opening. A serious imbalance between aggregate supply and demand does not in itself reflect deficiencies in the allocation process of a single market. It may be argued, in fact, that the burden on this process is less when such imbalances are evident. When jobs are plentiful, *ceteris paribus*, workers will have little trouble finding them; when unemployment is high, jobs will not stay open for long. In other words, when there are many more jobs than workers seeking to fill them, or vice versa, market information does not *need* to be disseminated as efficiently or thoroughly to effect all feasible match-ups as when market-wide supply and demand are roughly in balance. The question, then, is one of allocating labor *among* markets rather than *within* a given market. The difference is between the needs of the industry as a whole in a given area and those of the individual firm.

The process of matching worker to jobs is of special importance in construction. The reason, of course, is the intermittency of employment to which the industry's work force is characteristically subject. The constant disassembling and reassembling of work groups, and the pattern of repeated shifts from one employer to another simply means that a large number of separate matchings must take place. Given the structural dispersion of the industry, however, the task of informing workers about work opportunities in the area and/or that of making employers aware of the availability of certain workers is a prodigious one. In short, the varying needs of individual employers, as well as the large number of employers to begin with, combine to place extraordinary demands on the allocative functions of the construction labor market.

In this context, there would appear to be a call for some central clearinghouse of job market information, one in which both workers and employers could register their needs. As we shall see, there is no such system on the homebuilding markets surveyed. In the union sector, those arrangements commonly termed "hiring halls" serve an allocative purpose, although it is frequently argued that their central intent is to maintain union control over labor supply. In this chapter, we propose to examine the job matching process in homebuilding and to compare it with that of the union sector, in order to find some basis for evaluating its performance in residential construction.

HIRING IN RESIDENTIAL CONSTRUCTION: ERIE COUNTY

The hiring process in residential construction, and presumably in nonunion construction generally, is predominantly an informal one. Most labor market information is transmitted through a "grapevine" involving workers, builders, subcontractors, and material suppliers. The builder seeking a worker is apt to solicit applicants from his present employees, contact workers he has employed in the past, or ask his subcontractors or even his competitors if they know of anyone available. With the exception of newspaper advertisements, the most commonly used methods of finding workers do not involve any formal or systematic search. Unlike the union employer, the nonunion homebuilder and his subcontractors do not have access to a single agency through which their job openings can be made known to most prospective employees in the local labor market, much less those in distant areas. Occasionally, the local builders association is used informally by employers as well as workers as a source of job market information, but this practice is relatively infrequent. A few calls a month appear to be the norm, where it was mentioned at all.

The process by which one obtains employment is similar for the worker. If he is unemployed he is most likely to call on previous employers or ask his former co-workers if they know where work can be found. The experienced worker is likely to have developed enough connections to be able to find a job on his own without randomly visiting building sites, answering ads, or depending on the employment service. The young worker with marginal skills or experience may learn of job openings through friends or relatives, but he is more likely to seek work through more formal channels.

A somewhat broader picture of hiring may be obtained from the responses to the Erie County survey. Employers were asked to break down, in percentage terms, their new hires according to eight specified methods of recruitment. Since many firms were known to employ only a handful of workers and the method of hiring them is not normally a matter of record, it was deemed desirable to phrase the question in terms of general practice rather than in terms of current workers only or in terms of a particular time period. Furthermore, since respondents could not be expected to provide much more than rough estimates, the responses are shown in groups of rather broad categories (see Table 34).

The breakdown, crude as it is, does permit a few observations. Only one method of search was used by more than one-half the respondents—that of soliciting potential workers through present employees. A second method, newspaper ads, is also prominent as a major source of new hires. Unions and employment agencies are used infrequently. The other forms of hiring, utilized to varying degrees, are usually elements of the "grapevine" noted earlier. Previous employees and other employers are informed that openings exist, and some workers are hired

TABLE 34. *Homebuilding Industry*
Method of Hiring by Builders and Subcontractors
By Percent of Time Used
Erie County, New York, 1971

		Number of Respondents					
		Percent of Time Used					
Type of New Hire	Total [a]	0	1-20	21-40	41-60	61-80	81-100
Employed in the past	84	55	12	7	2	5	3
Recommended by current workers	84	32	17	12	10	5	8
Recommended by others	84	47	16	10	4	4	3
Not employed before but known by respondent	84	56	16	5	5	1	1
Hired through newspaper	84	48	5	8	7	8	8
Hired through union	84	78	3	—	1	2	—
Hired through employment agency	84	68	7	2	1	2	4
Hired "off the street"	84	59	13	2	3	3	4

Source: Erie County survey, 1971.

[a] Total includes all interviewed builders and subcontractors minus five no response.

upon direct application either having approached the firm at random or having heard that work was available there.

The picture in Table 34 is sharpened somewhat by the responses to another question on the survey: "If you needed a worker, what would you do *first* to get one?" Although the question was open-ended, the answers tended to fall rather neatly into specific categories (Table 35) and the relative importance of each category corresponds roughly with the pattern in Table 34. Newspaper advertisements and recommendations from present workers remain as the major avenues of hire. Employers are apparently somewhat less inclined to contract directly previous employees on their own. They are more apt to solicit referrals from acquaintances in the industry, although several respondents indicated that they maintain lists of previous applicants who were not hired at the time. Again, employment agencies are not extensively utilized.[111] In short, except to the extent that the newspapers are used to broadcast job openings, employers are generally quite content to rely on the grapevine to provide them with applicants. The operation of the grapevine is illustrated by the following comments:

—I would ask my men or lumber companies. They usually have an abundance of young fellows asking for work.

—You pass the word around that you need a worker and if someone is out of work they will call you.

—I'd call other contractors to find out if anyone worthwhile has been laid off.

—Go home and pray. Ask different people if someone is available. I have even called employment agencies when I'm desperate.

—I make a couple of phone calls to some friends or else my brother posts a notice in the plant.

The use of newspaper advertisements in the hiring process merits further discussion. Newspaper advertising is virtually unknown in the unionized sector of construction. In the homebuilding industry, however, it is apparently utilized to a much greater degree by open shop firms, as indicated in Tables 34 and 35. A number of respondents, however, expressed dissatisfaction with the results of their advertising. For example:

—In the past I have had disastrous results from people answering newspaper ads.

---

[111] When "employment agency" was mentioned, the agency referred to was invariably the Public Employment Service.

TABLE 35. *Homebuilding Industry*
Builders' and Subcontractors' First Move to Find Workers
Erie County, New York, 1971

First Move	Number of Respondents		
	Total	Builders	Subcontractors
Run advertisement	24	6	18
Ask present workers	22	9	13
Call other employers	16	8	8
Let out work generally	10	4	6
Call former workers	9	6	3
Call employment agency	6	—	6
Call vocational school	1	—	1
Total [a]	88	33	55

Source: Erie County survey, 1971.

[a] Total includes all interviewed builders and contractors minus one no response.

—I do occasionally run ads. It's a waste of time. I have never hired anyone through an ad.
—Put an ad in the paper, but then I get a lot of calls, but they don't have the experience to run the machinery.
—Would be reluctant [to run an ad] because you get people who say they're plumbers but aren't.
—Contact some other builders to see if anyone is floating. Paper last resort.

Some of the responses suggested that newspaper ads are used more to recruit relatively unskilled laborers. Although most of the employers indicated that their hiring procedures are the same regardless of the type of worker needed, a few did provide distinctions:

—Anyone could be a laborer. I use newspaper ads, employment agencies.
—A laborer you could hire through a newspaper—a carpenter never.
—If it's a truck driver I need, I advertise in the newspaper.
—Carpenter should be recommended, but sometimes hired through an ad.
—Skilled worker recommended by someone I know.

The thrust of the foregoing discussion is that there is no organized pool of workers from which prospective employers can systematically draw. When a firm needs a worker, it relies either upon a limited circle of acquaintances to transmit and absorb its labor market information, or upon newspaper ads which by their nature can perform no screening functions.

These same limitations are faced in many other industries, but certain features of construction render the difficulties of job matching more severe. Construction in general tends to have (1) smaller employing units which are even smaller in homebuilding; (2) a larger proportion of skilled workers; and (3) relatively unsteady employment patterns. As a result, construction does not exhibit a well-developed intrafirm labor market structure which enables the firm to fill jobs from within. Clark Kerr has distinguished the internal labor market as having "points of contact with the external market rather than interconnections along the line-up of jobs."[112] Between these points of contact there is no reliance on the external market. The volatility of building activity, high turnover of employment, skill requirements, and the smallness of the internal pool of workers, however, mean that construction firms must establish "interconnections all along the line-up of jobs," rather than just a limited number of "ports of entry." A builder is as likely to hire a carpenter from the outside as to promote one of his carpenter helpers.

Thus, when hiring the marginal worker, builders have no convenient screening device by which to assure minimal qualifications in their new workers, a function which is performed, however imperfectly, by unions in the organized sector. When asked whether they use "any minimum standards for hiring a new worker" or "invariably try him out" on the job, about three-fourths of the Erie County employers surveyed answered the latter. Those who did have minimum standards responded mostly in terms of age, experience, and recommendations. As a result, some workers are hired without the necessary competence, and then the process must begin again. Furthermore, it takes time for information about job openings to be disseminated throughout the industry, and available workers may learn about vacancies slowly, if at all.

---

[112] Clark Kerr, "The Balkanization of Labor Markets" in E. W. Bakke (ed.), *Labor Mobility and Economic Opportunity* (New York: Wiley, 1954), p. 102.

Effectiveness of the Hiring Procedures

A number of questions on the survey sought to elicit the views of employers in regard to the effectiveness of the matching process. First, respondents were asked: "Have you ever been in a position where you simply could not get the workers you needed?" Of the 90 firms which did employ on-site labor, 49 answered affirmatively. In most cases, the inability to recruit sufficient help required some change in the firm's normal operations. Table 36 details the kind of adjustments utilized.

Another approach to ascertaining employers' satisfaction with the matching process was to ask them about the potential usefulness of a more formal, structured system. The obvious direction in this regard would be toward a kind of labor exchange, operated much like the Employment Service or even a union hiring hall. Accordingly, employers were asked the following question: "Do you think it would be a good idea or not to have a system (perhaps operated by the Builders Association) through which workers and builders/subcontractors could exchange information on jobs open and jobs wanted?" The overwhelming sentiment was in favor of such a system. Of 90 respondents, 72 expressed interest. The reasons for this interest were generally related to the kinds of problems outlined above: (1) the lack of a screening mechansim; and (2) the need for quicker dissemination of labor market information for both employers and workers. The comments below are illustrative.

—Good from the worker's standpoint. [It's a] seasonal business; particular contractor will have a lot of work and some will not. This way a worker will have work all the time.

—You'd be able to pick up a man with special experience. Might be able to hire in a hurry if you had a pressing job.

—People looking for jobs don't make the right contact.

—At least you'd know if others are laying off, and if you needed workers you could hire them right away.

—Be like a labor bank. Get better personnel. Screened before you hired them.

—You could find specifically the worker who was needed and who was available. You wouldn't have to take somebody blind.

—Because you'd get better class of workers, if he took trouble to go in. Also could get his qualifications.

—It costs to advertise and many times in response to ads we hire three or four workers by telephone interview. Then they don't show up on

TABLE 36. *Homebuilding Industry*
Employer Adjustments to Manpower Shortages
Erie County, New York, 1971

Nature of Adjustment	Number of Employers
Worked overtime	17
Employer did work himself	9
Refused or lost jobs	8
Ran job longer	6
Contracted work out	3
Used prefabricated material	1
No answer	5
Total [a]	49

Source: Erie County survey, 1971.

[a] Total includes all interviewed builders and subcontractors minus 40 not applicable because they never experienced a shortage.

the job the next morning. And this just hasn't happened once, but has been repeated again and again. At least with a sort of clearing house, the men would be more interested in working and would probably show up.

The negative responses to this question were based on several factors. For some employers, the system's drawback was also its main virtue. Because such a system would facilitate mobility, they feared losing their own workers. Others generally mistrusted the Builders Association for reasons that were not clear. Still others were simply satisfied with the present procedures. Some specific comments follow:

—When I get a good guy, I don't want him running off to work for someone else.

—Our work is competitive. We're in competition with the builders to do their work.

—There would be too much pilfering of good workers.

—I shy away from organizations. The big guys would end up with the good men. I don't see where it would help. We could get what's left.

—You got to have freedom of choice of people and it's hard to get that from a central organization.

—It's not necessary. Jobs are gotten on a competitive bid basis and I know who to call.

Another factor which generated some concern over a centralized information exchange was the possibility of such a system facilitating union access to the workers. By and large, this concern was expressed more by association executives than by individual contractors. Although it is not immediately clear to the author how an information exchange (which is not the same as a hiring hall), listing at any one time only job-seeking workers or employers with openings, could be used effectively by the unions (even if they could obtain such lists—a questionable assumption itself), it is obvious that the fear of unionization might well constitute a barrier to any realistic effort to establish a systematic job market brokerage. Furthermore, it should be noted that any resistance based on this fear is likely to be greater in other segments of nonunion construction where, as observed earlier, there is more direct competition with the union sector.

Again, however, the overwhelming majority of the Erie County builders and subcontractors surveyed expressed approval in principle of a more structured system of hiring and recruitment. Approval in principle, however, is not the same as endorsement in practice. Thus, to learn more about the strength of employers' attitudes toward a formal employment exchange, subsequent questions were asked relating to the terms of expected usage and cost of this system.

As noted above, 72 of the 90 respondents thought that a formal hiring mechanism was a good idea. However, when asked whether they would have occasion to *use* such a system, only 62 answered affirmatively. Apparently ten employers felt that the arrangement would be a good idea for the other fellow. Presumably the other fellow would also pay for it.

The extent of an employer's willingness to subsidize an employment exchange probably provides the best indication of his feelings about the relative need for it. We first asked whether they would "be willing to make a *yearly* contribution [of an unspecified amount] to help run this system?" Here the number of affirmative answers dropped to 52 "yes" and one "maybe." A further question asked how much of an annual contribution they would be willing to make. The responses were as follows: 18 claimed $100; nine, $50; nine, $25; and 17 declined to specify the amount. Alternatively, 26 respondents would be willing to pay according to size and 14 more said that they would consider such an arrangement. Several employers, howover, would prefer to pay a set fee each time the system was *used.* This approach, of course, would discourage its use. Since only workers at the margin of demand would be hired this way, the system would probably fail to be self-supporting or fall into disrepute.

Thus far we have not distinguished among the different kinds of employers in the survey. It should be noted, however, that the builders, particularly those who employed on-site workers, were discernibly more favorable toward the establishment of an employment exchange than were the subcontractors. Thus, although 88 percent of the builders felt that such an exchange would be beneficial, only 72 percent of the subcontractors expressed this same sentiment (see Table 37), and whereas 73 percent of the builders were willing to pay for it, only 49 percent of the subcontractors were so willing.

There are a number of possible explanations of the relatively favorable reactions of builders to the idea of an employment exchange. First, virtually all of the builders surveyed are members of the Builders Association, whereas most of the subcontractors are not. Some of the negative comments by subcontractors explicitly expressed distrust for the Association. Second, the subcontractors already rely to a greater degree upon the more formal methods of recruitment. Although 18 percent of the builders mentioned newspapers as their first avenue of search and none mentioned employment agencies, the corresponding percentages for subcontractors were 33 and 11, respectively. Thus the subcontractors might be more inclined to view an employment exchange as a duplication of their usual methods and, therefore, not worth the cost. Finally, the builders may have perceived a greater need for some kind of *screening* device for job applicants. General contractors, as a rule, employ a larger proportion of unskilled and semiskilled workers and are apt to receive more applications from persons with limited skill and experience.

Employer Size and Hiring Practices

The results of the Erie County survey revealed some differences between large and small employers in regard to their practices and attitudes toward hiring. For the purposes of this discussion, large employers are defined as those with ten or more workers during the busy season (N = 32) and small employers as those with less than ten (N = 56). In general, the larger

TABLE 37. *Homebuilding Industry*
Builders' and Subcontractors' Responses
Concerning the Usefulness of a Formal Employment Exchange
Erie County, New York, 1971

	Builders					Subcontractors				
Question	Total	Yes	No	Not Sure	No Answer	Total	Yes	No	Not Sure	No Answer
Exchange a good idea?	33	29	4	—	—	58	41	14	2	1
Would you use it?	33	26	7	—	—	57	36	21	—	—
Would you contribute annually?	33	24	9	—	—	57	29	26	1	1
$100?	24[a]	8	10	6	—	29[a]	10	8	11	—
$50?	24[a]	15	3	6	—	29[a]	12	6	11	—
$25?	24[a]	18	—	6	—	29[a]	18	—	11	—
Would you contribute 1¢ per hour?	33	11	17	4	1	57	15	32	10	—

Source: Erie County survey, 1971.

[a] Limited to those who said they would or might contribute annually.

employers tended to place greater reliance on relatively formal methods of hire, specifically newspaper advertisements and employment agencies. These two sources were mentioned as the first avenue of search by one-half of the larger firms and only one-quarter of the smaller firms. Conversely, newspapers were *never* used by 63 percent of the smaller firms and 45 percent of the larger firms. The percentages on use of employment agencies were 86 and 71, respectively.[113]

Larger firms apparently experience more difficulty in finding the workers they need. In response to the question of whether the respondent ever has been unable to attract the necessary help, 48 percent of the smaller firms and 66 percent of the larger firms answered affirmatively. On the other hand, there was not as clear a difference between large and small firms with respect to their views on the desirability of a formal referral system run by the Builders Association. For example, about 56 percent of the small employers and 63 percent of the large ones expressed a willingness to contribute to such a system.[114]

In summation, present hiring procedures appear to operate somewhat more satisfactorily for smaller employers. Since their need is generally for one or two additional workers at a time, they are better able to fill openings by advertising them through informal channels. Nevertheless, the differences between small and large firms are probably less important that the similarity of their problems. A substantial proportion of both groups had had some difficulties in recruiting labor, and both were generally in favor of a more formal and systematic method of job matching.

HIRING PROCEDURES OUTSIDE ERIE COUNTY

Our interviews with the builders and Builders Association officials in other communities revealed, on the whole, that hiring methods in Erie County are typical of those used elsewhere. Many of the responses of Erie County employers were almost duplicated by spokesmen in other areas where homebuilding operates primarily on a nonunion basis. In this section we shall attempt to convey the attitudes and perceptions of those interviewed outside Erie County.

---

[113] Erie County survey, 1971.

[114] *Ibid.*

Rochester

According to the Executive Officer (E.O.) of the Rochester Builders Association, labor market information is largely a matter of word of mouth among the workers. Frequently, a worker will learn that building is being done at a certain location and will apply directly to subcontractors at the site. As in Buffalo, most Rochester builders do not employ their own crews. Occasionally, workers will come to the Association office to ask where work might be available. The E.O. felt that communication among workers is much more extensive than that among employers, although he did note that market information does tend to be exchanged among employers at social functions. Employment agencies are used rarely and newspaper advertising is used only as a last resort. Although the E.O. alluded to a chronic shortage of workers, he thought that the grapevine worked very well. When asked specifically about the desirability of an employment exchange, his response was that builders would be afraid to have the Association assume too much of a labor market role, largely because of a fear that centralization might lead to unionization.

A major builder in Rochester, by and large, corroborated the E.O.'s account, although he placed somewhat greater stress on the importance of newspaper advertising. Primarily, he said, word of mouth is relied upon to inform workers that jobs are available. Although he employed no on-site crews himself, he likened the recruitment procedure to that by which a builder seeks out subcontractors whose relationship to the builder, as noted in a previous chapter, may not be fundamentally different from that of an employee. Subcontractors are generally found through communication with other builders or materials suppliers.

Syracuse

The E.O. of the Syracuse Association also emphasized the informal nature of the hiring process. Occasionally, workers and employers seek each other through the Association itself, but not in any systematic way. Newspaper advertisements are used infrequently and responses to them are usually from young people who wish to learn a trade. The grapevine is the major source of information. In regard to the effectiveness of these methods, the E.O. felt that "the job gets done." However, he had no opinion concerning whether an employment exchange would be an improvement.

In Syracuse the author also spoke with three builders who do employ on-site labor to some degree. Two of them relied mostly on the grapevine, and both expressed some dissatisfaction with the results of newspaper advertisements. The third builder, who also performs a considerable amount of nonresidential work, made more extensive use of newspaper advertising. Newspapers, however, were felt to be much better for recruiting helpers than for journeymen, since hiring through advertisements requires a great deal of screening "to weed out the incompetents." This builder also hired workers on direct referral and a small number through the unemployment service. He reported no difficulty in obtaining sufficient help, although it is noteworthy that he is a large employer by construction standards and thus has more flexibility in allocating his manpower than does the average builder.

Albany

The E.O. in Albany was new at the time of our visit and had little knowledge of labor market operations. We talked briefly, however, with two builders about their hiring procedures. One spoke principally of newspaper advertising and secondarily of the State Employment Service and appeared to be reasonably satisfied with the results. The other builder, however, felt that advertisements generally produce an inferior worker and he preferred to rely upon word-of-mouth to generate applicants. His opinion was that a central referral system would improve the process.

Boston

The E.O. reported that builders relied chiefly upon informal information channels to obtain both workers and subcontractors. Newspapers are used but not extensively, whereas some trades are recruited through vocational schools. Occasionally, there is an exchange of information through the Association itself, primarily in regard to supervisory personnel. When asked to evaluate the process as it currently works, the E.O. responded that "it hasn't been brought up at meetings, so perhaps it's not a problem."

Providence

This E.O. also spoke of hiring in terms of newspaper advertising and word-of-mouth. His opinion was that the process is not efficient, and he would like to see a greater role for the Association in this regard. It is worth mentioning here that an agreement between the Association and the Carpenters Union, covering several Association members, makes no reference to hiring or union referrals.

Cleveland

The situation in Cleveland is different from those in the other areas. Historically, Cleveland's residential construction industry has been tightly organized, although the unions have gradually been losing control as residential building has moved farther and farther from the city. Furthermore, for the past several years construction generally has suffered a great deal of unemployment. As a result, many builders have lists of previous employees upon whom they can call to fill openings. There is also, according to the E.O. in Cleveland, a well-developed grapevine among both workers and employers. Some employers even "loan out" workers with the understanding that they will be returned if needed. As for the hiring hall, the E.O. reported that it is used by builders when jobs are plentiful, but in recent years it has not been needed. He noted that, in any event, the hall is used only as a last resort, since many builders feel that only workers with limited talent have to rely on it to find work. Newspaper advertising is used infrequently, and then only by nonunion builders. It must be emphasized, however, that the problem of matching jobs and workers is seldom severe under conditions of a substantial manpower surplus.

Other Areas

The interviews by the Industrial Research Unit with builders and Association spokesmen in Baltimore, Denver, Indianapolis, and Philadelphia, concentrated on the desirability of a central information exchange. The reactions there were predominantly negative. The factor mentioned most frequently was the threat of unionization. There was also a feeling on the part of some respondents that any workers hired through such a system would be of questionable quality. The negative responses tended to be stronger among Association personnel than among builders, although the handful of builders interviewed in these areas was generally less receptive to the idea than those in Erie County.

Conclusion

Whether informal methods of recruitment are adequate to the task of efficiently matching jobs and workers will obviously depend on one's standard of efficiency. In the following section, hiring procedures in the union sector, where central referral services are maintained in most markets, are explored for the purpose of providing a standard of comparison with the homebuilding industry. Hiring procedures in union construction, to the extent that they center on the union hall, serve to illustrate both the advantages and possible pitfalls of a centralized information system. An evaluation of the hiring process in homebuilding, in light of the comparative experience in the union sector, is reserved for the concluding section of the chapter.

REFERRAL SERVICES IN UNION CONSTRUCTION

Union referrals may operate in a variety of ways. In general, an employer who cannot find workers by other methods informs the union that he has an opening. The union may or may not have an established procedure for ordering its referrals and, therefore, may dispatch workers according to either a fixed registration list or the personal priorities of union officials. Federal law requires that referrals not discriminate between union and nonunion workers. If the local union knows of no unemployed workers, it will usually inform other locals that there are job openings in its area. Although unions are unlikely to welcome nonmembers with open arms, when some of its members are unemployed, its services are readily available to nonunion members when the membership is fully employed. This system goes far to ensure that vacancies will not persist for long when there are workers available in the market. In short, the employer is usually able to secure workers from the union in relatively short order, although the quality of the worker is likely to decline as the market tightens.

Despite the widespread popular notion that union hiring halls have historically been used as an instrument of job control, there has been surprisingly little empirical evidence of the hiring process in practice. Although it is clear that most building trades unions have always provided a job referral service for both its members and contractors, it does not follow that this service itself constituted a method for regulating the hiring process. Philip Ross, in a recent study of the origins of the construction hiring hall, carefully distinguishes a referral service from a system of control:

> A union hiring hall is a union-administered job-referral system whose referrals enjoy advantages over all other job applicants. . . . [W]hat is essential to all hiring halls as defined here is employer agreement or acquiescence in according job preference; it is both historically inaccurate and analytically useless to call union referral services "hiring halls" where there is no job preference.[115]

Prior to the enactment of the Taft-Hartley Act in 1947, there were few instances in which employers were *required* to utilize union facilities when seeking new workers. Control over hiring was exerted, not through a hiring hall, but through the closed shop, which was enforced at the job site by union stewards and business agents. "Employers were free to hire whom they wished, just as long as they were members of the union."[116] Union members were free to seek work on their own, and those who did utilize the union hall for information as to job opportunities had no contractual or practical advantage over those who applied directly at the site.[117]

Since 1947, however, the use and the application of the hiring hall, as defined by Ross, has expanded. With the outlawing of the closed shop by Taft-Hartley, and the later ill-fated attack on discriminatory hiring procedures by the NLRB,[118] unions

---

[115] Philip Ross, "Origin of the Hiring Hall in Construction," *Industrial Relations*, Vol. XI (October 1972), p. 366.

[116] *Ibid.*, p. 368.

[117] Ross does note certain earlier practices which had the effect of conferring advantages on union referrals, including the permit system, waiting lists, and work sharing rules. It appears, however, that none of these practices was particularly widespread. *Ibid.*, pp. 373-375.

[118] See *Mountain Pacific Chapter of the Associated General Contractors*, 119 N.L.R.B. 883 (1957). This case held that exclusive referral systems would be *per se* violations of the law unless they explicitly provided: (1) referrals shall be made on a nondiscriminatory basis; (2) the employer retains the right to reject any applicant referred by the union; and (3) the parties post conspicuously all provisions relating to the hiring system including the above two safeguards. Under another ruling, violations would be remedied by reimbursement of *all* dues and fees collected by the union since six months prior to the beginning of the litigation. Local 231, United Association

began to adopt formal hiring hall arrangements. In some cases, the formal changes did nothing to alter actual practice based on the closed shop. In others, there were elaborate rules establishing priority preferences in union referrals, based on previous employment, years of experience, local residence, and demonstrated qualifications.[119] At the present time, however, there is little empirical evidence regarding how the formal arrangements operate in practice. As Ross concludes,

> It is difficult to classify these exclusive hiring halls because the imposition of formal procedures upon a wide variety of local practices has produced situations where the formal hiring procedure exists next to a shadow system of job allocation which may or may not correspond to the written rules. . . . The most that we can say is that the degree of enforcement of the written rules depends primarily upon the degree of member unemployment and the inclination of union leaders to exploit the internal political opportunities offered by the hiring hall. . . . On the surface, all hiring halls conform to the legal requirement forbidding discrimination against nonmembers, however, any conflict between hiring hall rules and the attainment of union goals is inevitably resolved in favor of the latter.[120]

What little information there is suggests that practices vary greatly from craft to craft and from area to area. On the formal level, a recent study by the U.S. Department of Labor attests to this variety. The study surveyed the formal hiring procedures specified in agreements covering one thousand workers or more—291 contracts in all. Of these agreements, 132 (45 percent) provided for "exclusive" work referrals, meaning that the employer is required to use the union hall as the sole source of workers; 98 others were "nonexclusive" in that employers were allowed to hire from other sources; and 61 had no referral provisions at all.[121] Exclusive referral systems were most common among electricians, operating engineers, and plumbers, and least common among bricklayers, carpenters, ironworkers,

---

[Brown-Olds], 115 N.L.R.B. 594 (1956). The *Mountain Pacific* doctrine was held invalid by the U.S. Supreme Court in *Local 357, Teamsters Union* v. *NLRB*, 365 U.S. 667 (1961).

119 These priorities are explicitly sanctioned in section 8(f) of the Taft-Hartley Act under the 1959 amendments.

120 Ross, *op. cit.*, pp. 378-379.

121 U.S. Department of Labor, Labor-Management Services Administration, *Exclusive Union Work Referral Systems in the Building Trades* (Washington, D.C.: Government Printing Office, 1970), p. 25.

painters, and teamsters. Of the 82 exclusive systems selected for further analysis, most provided explicitly for nondiscriminatory referrals and the employer's right to reject applicants; however, 51 specified a procedure by which workers would be dispatched according to the date of registration and the priority schedule. There were substantial variations in both procedures and priority classifications among the several systems.

A somewhat earlier study in California by Gordon W. Bertram also found a variety of formal hiring arrangements. One agreement allowed the contractor to bypass the hall whenever he was hiring in his own local district.[122] In most other agreements, the contractor had the right to request a specific worker from the hall, although the extent to which this request would be honored depended in varying ways on certain eligibility standards (e.g., residency and employment experience in the area) to be applied to the worker in question. The general emphasis on locality preference found by Bertram is consistent with Ross's observation that most complaints to the NLRB alleging discriminatory hiring came from travelers.[123] Bertram's analysis concluded that, on the whole, hiring practices did conform to the procedures set forth in union agreements.[124]

Despite their variation, "exclusive" hiring hall arrangements frequently allow the union employer some flexibility in his hiring decisions. In many instances he is free to rehire the workers he employed in the past, to hire a certain number of "key" employees without using the hall, to specify a need for a worker with certain qualifications, and to reject applicants he does not deem suitable. At the same time, the formal distinction between an exclusive and nonexclusive hiring hall can be misleading. One critical responsibility of a local union is to preserve work for its members. When demand is slack, employers will be under pressure to hire local members regardless of contractual language and will frequently solicit workers "voluntarily" through the union, rather than hire nonmembers or members of other locals. When jobs are plentiful, an exclusive hiring system may not be rigidly enforced by the union, and members may be al-

---

[122] Gordon W. Bertram, *Consolidated Bargaining in California Construction*, Monograph Series No. 12 (Los Angeles: University of California, Institute of Industrial Relations, 1966), Ch. VI, p. 88.

[123] Ross, *op. cit.*, p. 376.

[124] Bertram, *op. cit.*, pp. 89-92.

lowed to seek work on their own regardless of their position on the registration list.

From the employer's standpoint, however, the rigidity with which formal rules are adhered to may seem somewhat academic. If either contractual arrangements or informal pressures prevent him from hiring the individual worker he wants, because of either the worker's lack of union membership or low place on the union registration list, the employer is unlikely to regard the system as optimal. The nonunion homebuilder, whatever his difficulties in locating workers, still has virtually complete discretion in choosing among whatever applicants present themselves. In this context, therefore, Ross's distinction between a referral service and a hiring hall is a critical one.

Just as the restrictions imposed by the hiring hall highlight the flexibility enjoyed by nonunion builders, its value as a central source of information for both workers and employers points to the inefficiencies inherent in a haphazard job matching process, especially in a large labor market where employment opportunities are very dispersed. This is not the place to explore whether access to *needed* workers is worth the price paid by the union contractor in terms of the possibility of his being refused the specific workers he *desires*. The focus here is on the nonunion builder, and the difference is that he would not necessarily have to pay that price. At the same time, he would be reasonably assured that if he were unable to recruit help, it would be because there was simply no help available, and not because his message was not being delivered to the right people.

The hiring hall, therefore, does constitute a limitation on the employer's freedom to hire whom he pleases. As Bertram observed, "The employment referral standards . . . represent a compromise between contractor and union interest."[125] Still, the history of union hiring halls does not suggest that the evolution from a simple referral service is an inexorable one. The hiring hall came about in response to specific legal and institutional stimuli. These stimuli are not applicable to nonunion residential construction. There is no apparent reason that a referral service in homebuilding could not be operated much like any employment exchange, leaving the hiring decision completely in the hands of the employer and merely providing him with systematic information on which to base that decision.

[125] *Ibid.*, p. 92.

CONCLUDING REMARKS

The discussion in the foregoing pages suggests that there is a discernible parallel between hiring in nonunion residential construction and the amount of hiring in the union sector which is not done through an exclusive referral system. In both sectors, a good deal of labor market information is transmitted informally among both workers and employers. Some hiring is also done by direct application at the job site, although this phenomenon is probably more prevalent in nonresidential construction because of the generally larger size of commercial projects. In both cases, workers are sometimes contacted directly by previous employers. The difference, however, is that in homebuilding there is no central agency to supplement these methods of recruitment.

At first, one might be tempted to regard formal employment agencies as the nonunion counterpart to the hiring hall. Our inquiries in both Erie County and elsewhere, however, reveal that such agencies are seldom used by either builders or subcontractors. In a sense, the use of newspaper advertising can be construed as the real counterpart to union referral systems. For some builders, advertisements are the first avenue of search, just as some union contractors automatically turn to the hiring hall when workers are needed. For others, advertisements are a last resort to be used only when informal methods of recruitment fail. It is noteworthy that a major criticism of advertisements—they produce an inferior worker who cannot get a job any other way—is much the same as laments voiced by some union employers in regard to the hiring hall. In both instances there is the complaint that workers are not adequately screened before they are allowed to complete an application.

The question of screening is only part of the broader operational test of hiring procedures, namely, the efficiency or perhaps speed with which an unemployed worker is matched with a job opening. In other words, inadequate screening introduces delay into the process of filling an available job with a worker possessing the requisite skills. The generally favorable response of Erie County employers to the prospect of a structured employment exchange suggests that existing procedures have considerable room for expedition. Not only do these procedures fail to provide any screening function at all, but they also take time. Jobs may remain open for days or weeks before information concerning their availability reaches a prospective applicant. With newspaper advertisements, employers may waste time trying out applicants who prove to be unqualified. In an industry that is as decentralized as construction, the usefulness of an agency through which information is given and sought, and which is familiar and accessible to labor market participants, would appear to be obvious.

It is by this standard that a central referral service becomes attractive. The union contractor, however dissatisfied he might be with referrals from the hall, can nevertheless generally secure a worker with a single telephone call. Moreover, he may in some circumstances specify that he needs someone with particular skills and also may be able to obtain *some* information on a worker's background. To be sure, the same information is probably available through informal channels, but these informal channels, upon which the homebuilder must rely entirely, are by their nature more likely to be haphazard and time consuming.

At the same time, the effectiveness of a formal employment exchange will depend upon the willingness of employers and workers to use it. The currently meager use of the state employment service does not offer much hope in this regard, but it may be that a system indigenous to the industry would be better accepted. Such a system would not be burdened by the image of the employment service as a "passive adjunct of the unemployment office" [126] which has in the past discouraged both job seekers and prospective employers from availing themselves of its services. Furthermore, it could be fashioned to accommodate the particular needs and problems of the industry and staffed by persons familiar with them. Most hiring would probably still be done informally, but it is the *marginal* hire which will determine the effectiveness of any procedure, and it seems clear that under many circumstances the marginal hire would be expedited by a more systematic exchange of labor market information.

Given the apparent interest of residential contractors in a formal employment exchange, the government might well explore ways to encourage and assist local Associations in establishing their own employment information systems. In particular, the U.S. Employment Service and state services could

---

126 "Report to the Secretary of Labor from the Employment Service Task Force," in R. A. Gordon (ed.), *Toward A Manpower Policy* (New York: Wiley, 1967), p. 147.

offer their technical expertise to Associations wishing to develop such procedures, but unfamiliar with the means of development. As with training, we do not envisage a permanent subsidy of the industry's operations; the function of matching workers and jobs is one for which the industry itself should be responsible. Yet the issue involves a real question of labor market efficiency, and to that extent it does constitute a social problem. Moreover, since the U.S. Employment Service should presumably be prepared to perform this function itself under its general mandate, there would appear to be an even greater justification for its merely helping the industry to get started.

It must be acknowledged that the above proposal is feasible only to the extent that the need for it is perceived by the parties involved. Furthermore, they must see it as an opportunity rather than as a threat, a perception which as we have seen is not universal. In this connection, therefore, we have admittedly made only a *prima facie* case. Certainly, any public expenditures should not be made without some further exploration of the magnitude and nature of the problem to which they would be directed. At the least, however, the evidence here suggests that such further exploration is warranted.

CHAPTER VI

Wages and Benefits

In the late 1960's and early 1970's, skyrocketing wage settlements in the construction industry commanded widespread public attention and concern. These settlements were disturbing both because of their cause, perceived largely as excessive institutional power wielded by the building trades unions, and because of their effects on the cost of construction itself and on wage bargaining in other sectors of the economy. Union wages in construction came under public control nearly five months before the general stabilization program was initiated in August 1971. Even after mandatory controls on most of the economy were relaxed in the winter of 1973, construction was one of the few sectors to remain subject to detailed governmental wage monitoring.

The matter of wage determination in union construction is an exceedingly complex one. For our purposes, however, the central issues are the relative levels and structures of wages in nonunion homebuilding and the union sector of the construction industry, and the implications of that relationship on the operation of residential labor markets. Therefore, determinants of union wages will be discussed only briefly, although the comparison with residential pay levels may provide some useful material for analyzing and evaluating compensation in the union sector. This chapter will assemble and examine available government data on construction wages, both union and nonunion, as well as the results of the Erie County survey and unpublished information obtained from the NAHB. Since union wages tend to be substantially higher than those in homebuilding, the major focus will be on the extent, behavior, and significance of this differential. Finally, it should be emphasized at the outset that the question is ultimately to be framed in terms of relative *total* compensation, and not simply the narrower and less meaningful concept of wage rates.

WAGE LEVELS IN THE UNION SECTOR

Since World War II, wages in construction have consistently risen at a faster rate than those in the rest of the economy. In 1947, gross average hourly earnings of production workers in contract construction were 26.2 percent higher than those in manufacturing and 36.3 percent higher than in all private nonagricultural industries. By 1972, these differentials had grown to 59.8 percent and 65.9 percent, respectively.[127] These figures, moreover, encompass both union and nonunion workers; the differentials are even greater when union establishments are considered separately. In 1969, *total* compensation including supplementary benefits by union subcontractors in construction was 52.6 percent higher than that by union establishments in manufacturing, whereas the difference in nonunion firms was only 15.6 percent.[128]

The patterns of wage rates in the various construction trades are too complex and too imperfectly understood to be discussed extensively here, but there is one aspect of the internal wage structure that bears mentioning. For many years, there has been a steady compression of the difference between union rates for skilled and unskilled workers. Between 1947 and 1971, for example, the journeyman-laborer differential fell from 52 percent to 27 percent.[129] In the more heavily unionized regions of the country, moreover, the differences are even lower (e.g., 17 percent in New England and 22 percent in the Middle Atlantic states).[130] As we have seen, skilled differentials tend to be greater in nonunion homebuilding.

The relatively high wage rates in construction are not uniformly translated into proportionately higher annual earnings due to the intermittent character of employment in the industry. A 1970 Bureau of Labor Statistics study found that median earnings for male year-round full-time union workers was $11,-

---

127 *Monthly Labor Review*, Vol. XCVI (June 1973), p. 98.

128 John Litsas, "What Subcontractors Pay Construction Workers," *Monthly Labor Review*, Vol. XCV (September 1972), p. 58.

129 U.S. Bureau of Labor Statistics, *Union Wages and Hours: Building Trades, 1971*, Bulletin 1747 (Washington, D.C.: Government Printing Office, 1972), p. 3; see generally Arthur Rose, "Wage Differentials in the Building Trades," *Monthly Labor Review*, Vol. XCII (October 1969).

130 U.S. Bureau of Labor Statistics, *Union Wages and Hours*, *op. cit.*, p. 14.

244, compared to $9,115 for all industries and $8,680 for manufacturing.[131] The differentials were thus 23 percent and 29 percent, respectively. Although the Bureau of Labor Statistics does not define a "year-round full-time" worker, it seems clear that the use of this concept serves to magnify the real difference in annual earnings between construction and most other industries. An earlier study for 1964, which referred to workers with earnings in all four quarters of the year and did not distinguish between union and nonunion workers, found substantially smaller differentials.[132]

Finally, an increasing proportion of construction worker compensation has come in the form of employer contributions to various trust funds for such benefits as health insurance, pensions, vacations, and supplementary unemployment payments. Between 1965 and the fourth quarter of 1972, these contributions rose from 7.7 percent of the hourly wage to 15.3 percent.[133] These payments, moreover, rose faster for unskilled workers than for skilled workers, thus accentuating the decline of skill differentials noted earlier. Table 38 summarizes recent data on wages and fund payments for selected trades.

Determinants of Compensation in the Union Sector

There has not been advanced yet an entirely satisfactory explanation of wage levels and movements in unionized construction. The relative importance of market conditions and institutional (union) power in construction, as elsewhere, will probably never be definitely resolved.[134] At the same time, there is evidence that some of the factors frequently adduced by union leaders do not sufficiently account for wage movements in con-

---

[131] U.S. Bureau of Labor Statistics, *Selected Earnings and Demographic Characteristics of Union Members, 1970,* Report No. 417 (Washington, D.C.: Government Printing Office, 1972), p. 16.

[132] U.S. Bureau of Labor Statistics, *Compensation, op. cit.,* p. 71.

[133] U.S. Department of Labor, Office of Information, News Release No. 73-33, February 9, 1973, p. 3.

[134] For an early study of this question, see Stephen P. Sobotka, "Union Influence on Wages: The Construction Industry," *Journal of Political Economy,* Vol. XLI (April 1953), pp. 127-143. Sobotka concluded that wages in five craft unions were 11 to 20 percent higher in 1939 than would have resulted purely from market forces.

TABLE 38. *Construction Industry*
Union Wage Rates and Employer Trust Fund Payments
In Building Trades in Cities of 100,000 Inhabitants or More
United States, January 2, 1973

Trade	Hourly Wage	Fund Payments	Fund Payments as Percent of Wage
All trades	$7.51	$1.15	15.3
Bricklayers	8.23	1.23	14.9
Building laborers	5.79	1.03	17.8
Carpenters	7.67	1.27	16.6
Electricians	8.47	0.89	10.5
Painters	7.24	0.86	11.9
Plasterers	7.69	1.10	14.3
Plumbers	8.39	1.76	21.0

Source: U.S. Department of Labor, Office of Information, News Release No. 73-33, February 9, 1973.

struction in the past 25 years. Although it is true that conditions such as the high proportion of skilled workers in construction, the absence or rarity of certain fringe benefits common in other industries, the uncertainty of work opportunity, and the general arduousness and hazards of building work should be expected *ceteris paribus* to result in higher wages, it does not follow that the differences should be so great.

Furthermore, even to the extent that these factors constitute disadvantages to the construction worker, it is difficult to argue that the disadvantages have been exacerbated over time.[135] Although skill mix is difficult to measure precisely, it does not appear that construction skill levels have grown faster than those in, for example, manufacturing. Moreover, nonwage benefits have been growing at a faster pace in construction than elsewhere; we have already seen that construction seasonality, if anything, has declined slightly since World War II. Yet the differential between construction and other wages has not only

[135] For a discussion of available evidence on these points, see Howard G. Foster, "Wages in Construction: Examining the Arguments," *Industrial Relations*, Vol. XI (October 1972), pp. 336-349.

been maintained during this period, but has steadily expanded. In short, even if the factors mentioned justify higher wages in construction, they do not explain the faster *rate* of wage increases in the industry.

The most systematic attempt to analyze recent wage trends in construction is by Mills.[136] Mills argues that the extraordinary settlements of the late 1960's cannot be attributed solely to union monopoly power in the absence of any evidence that union power had been growing during the decade. His explanation of the phenomenon rests on the interaction of three factors: (1) a major increase in the demand for construction, along with concomitant manpower shortages, in highly organized and price-inelastic sectors of the industry; (2) generally favorable economic conditions in the rest of the economy which resulted in a significant drain-off of manpower available to supplement the existing construction work force; and (3) the decentralized bargaining structure of the industry which allowed and encouraged the leapfrogging of wages across crafts and localities.

Although Mills's analysis is a welcome tonic to the devil theories often advanced by popular commentators,[137] there still remains a number of vexing questions. In the first place, compositional changes in demand are not unique to the 1960's or to the construction industry. It is not clear, moreover, why tight labor markets elsewhere should divert manpower *from* construction when construction wages were much higher from the start. Finally, even if some settlements were consistent with expansionary economic conditions, the very fact that those increases were able to be duplicated in other sectors and areas is itself a reflection of some nonmarket force. Although Mills is persuasive in demonstrating the lack of concrete evidence as to growing union power, the economic and structural factors he cites do not seem sufficient to explain settlements providing for annual increases reaching 20 percent by the time controls were imposed.

In fairness, it is much easier to fault any explanation of a complex phenomenon than to offer alternatives, and Mills has performed a real service by demonstrating how a confluence of

---

136 Mills, *op. cit.*, pp. 61-70.

137 See, for example, Thomas O'Hanlon, "The Unchecked Power of the Building Trades," *Fortune*, Vol. LXXCIII (December 1968).

events can initiate a wage spiral. Union power alone does not explain the developments of 1967-1971; otherwise, how does one account for the moderate increases throughout 1966? It seems clear, however, that collective bargaining in construction does impose an expansionary bias on wage movements. Mills himself concludes that ". . . had average hourly earnings in all industries risen at 3.2 percent per year [from 1950 to 1968] . . . construction earnings would have tended to rise at 3.9 percent per year, assuming that labor market conditions were unchanging."[138] To the extent that there is an interaction effect which goes beyond the sum of market and power factors, so that union power itself is exploited more aggressively during an economic expansion (and as a corollary is underexploited during periods of stability), the mechanics and causes of this effect remain obscure.

WAGES AND BENEFITS IN NONUNION CONSTRUCTION

Separate wage data for nonunion construction are not collected on a regular basis; therefore, the behavior of union-nonunion differentials over time is difficult to analyze. There is little doubt, however, that nonunion wages are now appreciably lower than those in the union sector, although there is some evidence to suggest that the differentials have not been on the same order of magnitude throughout recent history. For example, as noted in a previous chapter, the Bureau of Labor Statistics conducted a comprehensive study of union and nonunion wage rates in 1936.[139] Although the differentials then varied considerably by region and by craft, their order of magnitude was approximately 50 percent. The union rate was generally one and one-half times the nonunion rate. That period, of course, was one of extraordinarily high unemployment. When Haber and Levinson conducted their survey in a more prosperous time (1952), they concluded as follows:

> While no precise quantitative data were collected, the field survey . . . obtained a wide sample of personal evaluations from union, employer, and other industry representatives. . . . On the basis of information gathered in these interviews, there is little doubt that the 1936 differentials had narrowed greatly in all the cities surveyed and had com-

[138] Mills, *op. cit.*, p. 78.

[139] Edward P. Sanford, "Wage Rates and Hours of Labor in the Building Trades," *Monthly Labor Review*, Vol. XLV (August 1937), pp. 281-300).

pletely disappeared in some. Wherever differentials existed, they rarely were judged to exceed 10 percent, on the average, though the nonunion range of rates was much greater.[140]

In more recent years, there have been scattered government studies which, if the Haber and Levinson observations are representative of the early 1950's, demonstrate that nonunion wages have not risen as rapidly as those negotiated by the building trades unions. According to one Bureau of Labor Statistics publication, the average straight-time pay in 1965 of building construction establishments in which a majority of workers were covered by union agreements was $4.18, compared to $2.79 for establishments in which a minority of workers were covered.[141] In 1969, according to another Bureau of Labor Statistics survey, employees of unionized special trades contractors were paid an average of $5.43 compared to $3.26 for their nonunion counterparts.[142] In both studies, moreover, the *proportion* of total compensation paid in the form of straight-time wages was less for the union firms, so that the differentials in total compensation were greater than those for wages alone. It may be noted that the union-nonunion differences were of the same order of magnitude as that found by the Bureau of Labor Statistics in 1936.

It does not appear that the differences in wage rates are completely balanced by better work opportunities in the nonunion sector. Table 39 summarizes some of the relevant results of the Bureau of Labor Statistics survey of annual earnings in 1970. Not only are the median annual earnings of male union workers in construction appreciably higher than those of nonunion workers, but the ratios are substantially greater than those for most other industries and occupations. For example, the ratio for male blue collar workers as a whole was 1.21, and 0.90 for manufacturing.[143] At the same time, the data are com-

---

140 Haber and Levinson, *op. cit.*, pp. 222-223.

141 U.S. Bureau of Labor Statistics, *Compensation*, *op. cit.*, p. 85.

142 U.S. Bureau of Labor Statistics, *Employee Compensation and Payroll Hours: Construction-Special Trades Contractors, 1969*, Report no. 413 (Washington, D.C.: Government Printing Office, 1972), p. 3.

143 U.S. Bureau of Labor Statistics, *Selected Earnings*, *op. cit.*, pp. 13, 16.

TABLE 39. *Construction Industry*
Median Earnings of Year-round Full-time
Union and Nonunion Male Workers
In the Building Trades
United States, 1970

Category	Union Median	Nonunion Median	Union/Nonunion Ratio
Construction industry	$11,244	$7,619	1.48
Carpenters [a]	10,274	6,897	1.49
Other craftsmen [a]	11,223	7,853	1.43
Laborers [a]	8,693	5,128	1.70

Source: U.S. Bureau of Labor Statistics, *Selected Earnings and Demographic Characteristics of Union Members, 1970,* Report No. 417 (Washington, D.C.: Government Printing Office, 1972), Tables 6, 7, pp. 13, 16.

[a] Occupational categories not limited to workers in the construction industry.

promised by the limitations to year-round full-time workers, since the average part-year worker in union construction may work fewer hours than his nonunion counterpart. Nevertheless, in conjunction with the other studies, it does reflect a substantial differential between union and nonunion earnings.

The results of the annual earnings survey reflect another important feature of nonunion wages. In general, nonunion rates tend to be more dispersed than those paid by union employers. For example, almost 32 percent of the male nonunion workers earned either more than $15,000 or less than $5,000 in 1970, compared to just over 20 percent of male union workers. Table 40 illustrates the relative dispersion of union and nonunion earnings in 1970. Although these distributions reflect differences between markets, as well as within markets, we shall present further evidence in the following section which depict nonunion variation in the wage rates of a single labor market.

Unfortunately, none of the aforementioned studies makes any distinction between homebuilding and other nonunion construction. Another Bureau of Labor Statistics survey canvassed over 4,200 general and specialty contractors engaged in single-family building in 1969 and found an average wage rate of $3.94 compared to an average union rate of $5.14, but the entire union-nonunion differential is not reflected in these figures since nearly

TABLE 40. *Construction Industry*
Percent Distribution of Earnings of Year-round, Full-time Union and Nonunion Male Workers
United States, 1970

Earnings	Union	Nonunion
Under 3,000	1.8	5.3
3,000-4,999	1.6	14.5
5,000-6,999	9.0	25.0
7,000-9,999	26.7	25.5
10,000-14,999	43.9	17.7
15,000 and over	17.0	12.1
Total	100.0	100.0

Source: U.S. Bureau of Labor Statistics, *Selected Earnings and Demographic Characteristics of Union Members, 1970,* Report No. 417 (Washington, D.C.: Government Printing Office, 1972), p. 16.

Note: Percents may not add to 100.0 due to rounding.

one-third of the employers in the Bureau of Labor Statistics sample were themselves unionized.[144] Still, the results of this study (Table 41) reveal two significant points to which we shall have occasion to return. First, the percentage differentials between the homebuilding and average union rates were greatest for unskilled workers.[145] Second, the differences in average rates reflect not only spreads in the wages of individual crafts, but also the fact that the homebuilding work force is more heavily comprised of laborers and helpers. Not only are the individual wage rates lower in homebuilding, especially for unskilled workers, but the weights serve to increase the average differential. These weights are in part of a function of the greater flexibility enjoyed by nonunion employers in assigning workers to various tasks without being hampered by jurisdictional rules.

The lack of comparability among the various studies described in this section precludes any rigorous analysis of the behavior

144 U.S. Bureau of Labor Statistics, *Labor and Material Requirements, op. cit.*, p. 15.

145 This finding is consistent with the relative occupational differences in annual earnings shown in Table 38, p. 129.

TABLE 41. *Homebuilding Industry*
Average Hourly Earnings and Average Hourly Union Wage Rates For New Private Single-Family Houses, by Occupation United States, 1969

Occupation	Average Hourly Earnings	Average Union Hourly Wage Rate [a]
Carpenter	4.11	5.35
Painter	3.99	5.01
Bricklayer	4.76	5.63
Plumber	4.85	5.73
Cement finisher	4.53	5.12
Electrician	4.86	5.57
Plasterer	5.02	5.34
Sheet metal worker	4.60	5.48
Roofer	4.78	5.11
Operating engineer	4.36	—
Tile setter	4.14	5.25
Soft floor layer	4.68	—
Laborer	2.93	—
Helper and tender	2.56	4.05
Truck driver	3.54	—
All occupations	3.94	5.14

Source: U.S. Bureau of Labor Statistics, *Labor and Material Requirements for Construction of Private Single-Family Houses,* Bulletin 1755 (Washington, D.C.: Government Printing Office, 1972), p. 15.

[a] These figures cover all building trades in both residential and nonresidential construction as of July 1, 1969.

of union-nonunion differentials over time, although it is usually assumed that since nonunion wages are more susceptible to changes in market conditions, the tendency is for the differentials to expand when unemployment is high and contract when unemployment is low. At the same time, the level of construction activity does not behave the same in all sectors: the generally buoyant state of most construction in the late 1960's contrasted with a virtual depression in the residential sector. Without more comprehensive data on relative wage rates *and* the extent of unionization over time, it is impossible to reach firm conclusions on the complex relationship between market forces and wage differentials, although it is difficult to deny that some such relationship exists.

Residential Wage Rates and Benefits in Erie County

The sensitivity of union-nonunion differentials to labor market conditions becomes especially significant in light of two developments in Erie County since 1969. The first is the explosive pattern of wage increases negotiated in that year by most of the major building trades unions. The magnitude of these settlements is illustrated in Table 42, which shows total compensation for seven trades as of July 1, 1968 and April 3, 1972.

As Table 42 indicates, union wage scales were rising on the order of 18 percent per year. At the same time, labor markets in the Buffalo area were beginning to loosen after an eight-year period of almost continuously falling unemployment. In just two years, the overall unemployment rate for the Buffalo SMSA jumped from an annual average of 3.7 percent in 1969 to 7.8 percent in 1971, virtually wiping out the gains of the 1960's. Although construction unemployment is invariably higher than the all-industry average, the behavior of the two series is very similar. In short, since 1969, there have been more and more building tradesmen who, because of unemployment, have been unable to enjoy the full measure of their unions' bargaining gains, and consequently less and less pressure on nonunion employers to compete with the union scales.

The precise magnitude of the union-residential differentials is difficult to estimate for a variety of reasons. In the first place, the number of workers covered by the present survey was small for some of the crafts included; therefore, the findings cannot be represented as anything more than suggestive of the prevailing wages paid. Second, the range of nonunion rates is very wide, so that there are problems in comparing nonunion averages with the relatively fixed and uniform union scales. Third, there are differences in the ways in which various employers regard certain occupational definitions. A worker performing a specified family of tasks may be a carpenter to one employer, a helper to another, and a laborer to still another. Finally, workers in nonunion construction may perform tasks

TABLE 42. *Construction Industry*
Union Hourly Wages and Benefits for Selected Trades
Buffalo, New York, July 1, 1968 and April 3, 1972

Trade	Hourly Wages and Benefits		
	July 1, 1968	April 3, 1972	Percent Change
Bricklayers	6.07	9.67	59.3
Carpenters	5.715	9.48	65.9
Electricians	6.57	10.00	52.2
Painters	5.45	8.43	54.7
Plasterers	6.06	9.41	55.3
Plumbers	5.98	9.78	63.5
Laborers	4.785	8.635	80.4

Source: U.S. Bureau of Labor Statistics, *Union Wages and Hours: Building Trades, July 1, 1968,* Bulletin 1621 (Washington, D.C.: Government Printing Office, 1969), p. 20; U.S. Department of Labor, Office of Information, News Release 72-325, May 25, 1972, p. 4.

associated with more than one craft; this raises obvious difficulties in comparing them with union workers.

Accordingly, we shall not attempt to present pinpoint estimates of residential wages. Rather, we shall present the rates reported by the respondents in terms of several rather wide ranges, and relate them to union scales prevailing at about the same time. Although this approach is not as intellectually satisfying as a rigorous statistical analysis of wage patterns, it nevertheless illustrates forcefully the order of the union-residential differentials which have developed in Erie County. As will be seen below, these differentials are not atypical of those found elsewhere, and if anything, they understate actual differences in total compensation.

The wage rates reported by Erie County employers are summarized in Table 43. These figures represent responses to the question: "What is the *average* straight-time hourly wage of each of the types of workers you employ (as of 1971)?" It is clear from the table that the vast preponderance of workers in residential construction earn significantly less on an hourly basis than their counterparts in the union sector. Again, there are apparent discrepancies in the occupational definitions used by some respondents. It is difficult to believe, for example, that

TABLE 43. *Hombuilding Industry*
Residential Hourly Wages and Union Scales Reported by Employers, by Craft
Erie County, New York, July 1, 1971

Craft	Number of Employers Employing Craft	Number of Employees Employed	Number of Respondents — Wage Rate: Under 2.51	2.51-3.50	3.51-4.50	4.51-5.50	Over 5.50	Weighted Average	Union Scale
Carpenter	41	250	—	3	14	20	4	4.87	7.63
Cement finisher	12	24	—	—	3	5	4	5.03	8.56
Electrician	7	27	—	1	3	3	—	4.10	9.31
Machine operator	19	24	1	5	5	5	3	4.61	8.06[a]
Mason	5	5	—	—	1	4	—	4.90	8.11[b]
Painter	15	60	—	4	8	2	1	4.26	6.27
Plumber	9	43	—	3	2	3	1	4.67	8.88
Sheet metal worker	8	41	—	1	5	2	—	4.24[c]	8.50
Helper	29	85	2	21	6	—	—	3.32	—
Laborer	40	159	7	24	8	—	1	3.32	7.13

Source: Erie County survey, 1971; U.S. Bureau of Labor Statistics, *Union Wages and Hours: Building Trades, 1971* (Washington, D.C.: Government Printing Office, 1972), p. 20.

[a] Rate is different for various machines. Figure is for trench machines.

[b] Bricklayers only. Residential rates include blockmasons.

[c] Rate is for heating installer.

a fully qualified machine operator earns less than $2.50 or that a journeyman, electrician, or plumber earns less than $3.50. Nevertheless, the rates reported by the residential employers are so far below the union scale in most cases that it is probably safe to conclude that the average differential is at least as high as the 50 percent differential estimated by the Bureau of Labor Statistics in 1936.[146] The figures in Table 43, moreover, depict not only a substantial difference between union and residential rates, but also a much greater range of residential rates than that in the union sector, where the established wage is typically paid to most workers in the occupational category.[147]

The differences are even greater when nonwage benefits are taken into account. In the past decade or so, unions have directed an increasing proportion of their negotiated gains into benefit funds covering health insurance, pensions, vacations, and supplementary unemployment payments. By 1971, hourly contributions to these various funds had reached almost 14 percent of the basic wage for all crafts on an annual average. The magnitude of these payments in Erie County is illustrated in Table 44.

Residential employers in Erie County provide few supplementary benefits aside from those required by law (social security, unemployment insurance, etc.). To be sure, a number of benefits common to many industries, such as holidays, vacations, and severance pay, are relatively rare in any branch of construction. It is nevertheless apparent from the responses to our survey that the proportion of total compensation paid in forms other than wages is a great deal less in the residential nonunion sector. The various benefits on which information was solicited are discussed in turn below.

Health Insurance. Of 88 employers who responded to this question, 49 (56 percent) indicated that they provided some form of health insurance to their employees, and an additional three

---

[146] Sanford, *loc. cit.*

[147] Union scales actually represent the *minimum* allowable rate, and in tight labor markets employers may agree to pay some workers more than the minimum or to guarantee a certain amount of overtime. On the other hand, workers may agree to work overtime at straight-time pay in violation of the union contract. On the latter point, see Barry D. Welchel, "Informal Bargaining in Construction," *Industrial Relations*, Vol. X (February 1971), pp. 105-109. On the whole, however, it is doubtful that these practices produce an actual range of wage rates comparable to that in nonunion construction.

TABLE 44. *Construction Industry*
Hourly Contributions to Union Benefit Funds, by Craft
Buffalo, New York, July 1, 1971

Craft	Benefits	Wage Rate	Benefits as Percent of Wage
Bricklayer	1.55	8.11	19.1
Carpenter	1.85	7.63	24.2
Cement finisher	0.75	8.56	8.8
Electrician	0.59	9.31	6.3
Machine operator	1.30	8.06	16.1
Painter	1.175	6.275	18.7
Pipefitter	1.18	8.76	13.5
Plumber	0.70	8.88	7.9
Roofer	1.95	6.95	28.0
Sheet metal worker	1.20	8.50	14.1
Laborer	1.50	7.135	21.0

Source: U.S. Bureau of Labor Statistics, *Union Wages and Hours: Building Trades, 1971*, Bulletin 1747 (Washington, D.C.: Government Printing Office, 1972), p. 20.

respondents provided such benefits to some employees, usually supervisory personnel only. (It is possible, perhaps likely, that some of the 49 in fact provided insurance for only a fraction of their work force but failed to so signify.) Of those employers furnishing insurance, 37 pay all of the premiums.

Pensions and Life Insurance. Only 12 employers reported that they participated at all in a pension plan, and three more had such a plan for some employees. A total of 16 respondents provided life insurance.

Vacations. A total of 35 respondents allowed a paid vacation to their workers, twelve of whom specified that only some employees received it. Of the 35, 28 granted one week and 7 granted two weeks.

Holidays. Thirty-one employers had a practice of giving paid holidays and six more gave them to selected employees. (On the questions of both vacations and holidays, some respondents reported that these benefits were given to only one employee, the superintendent.) Of these 37, 16 allowed five days a year or less, 16 granted six or seven, and 5 did not specify.

Others. Employers said that the following benefits were given to some or all of their workers: use of car or truck (3); Christmas bonus (13); guaranteed 40-hour week or showup pay (5); and profit sharing (4).

The benefits cited by Erie County employers are summarized in Table 45. Perhaps the most noteworthy feature of the table is that except for health insurance, benefits very common in other industries, and even some almost universal in union construction (pensions and showup pay), are rare in homebuilding in Erie County. It is impossible, of course, to convert the costs of these benefits into precise hourly terms without more detailed information, but it nevertheless seems clear that averaged over the industry they amount to far less than union contractors' hourly contributions to their various trust funds.[148] The major benefit granted by some residential employers and not found in the union sector is the bonus or profit-sharing plan. The highest bonus figure mentioned was $500 a year, or about twenty-five cents per hour, and even this was a practice of only 17 of 88 respondents. Health insurance, assuming the quality of the plans is the same, appears to have roughly one-half the coverage in residential as in commercial construction where it is virtually universal to all crafts. The gap is even greater with respect to items such as pensions and private unemployment benefits of any kind. The vacations and holidays shown in Table 45 amount to about 12 paid days off for perhaps 30 percent of the work force or about four days on average. The approximate hourly cost, assuming a wage rate of $5.00 per hour, is eight cents.

The point here is merely to show that the union-nonunion wage differential, at least where nonunion *residential* construction is concerned, is not attributable to differences in the structure of compensation. If anything, the real differential appears to be greater than that suggested by wage rates alone. Another factor which may account for some of the difference is the availability of overtime. Unfortunately, there are no data on the extent of overtime in nonresidential construction. In Erie

---

[148] The Bureau of Labor Statistics study of 409 general building contractors in 1965 revealed that straight-time pay constituted 84.4 percent of total compensation in union construction and 89.4 percent in nonunion construction. U.S. Bureau of Labor Statistics, *Compensation, loc. cit.*

TABLE 45. *Homebuilding Industry*
Supplementary Benefits Paid by Employers
Erie County, New York, 1971

Type of Benefit	Total [a]	Number of Employers Granting	Number not Granting	Number Granting to Selected Employees
Health insurance	88	49	36	3
Pensions	88	12	73	3
Life insurance	88	16	72	—
Paid vacations	88	23	53	12
Paid holidays	88	31	51	6

Source: Erie County survey, 1971.

[a] Total includes all interviewed minus one no response.

County, we asked employers to "estimate how many hours of overtime your workers have averaged weekly over the past six months." Of 81 employers responding, 28 said none, 23 said one to six, and 30 said seven or more, with 14 hours a week the highest number mentioned.

Residential Wage Rates Outside Erie County

The union-residential differentials in Erie County are not unusual. In the majority of the cities visited only sketchy information on residential rates was obtained (generally limited to carpenters); however, these rates were invariably at least 25 percent below the union scale. In Syracuse, for example, one large builder paid his carpenters \$4.00 per hour, another up to \$5.25, and another between \$4.00 and \$6.00, compared to the union scale of \$7.82. In Albany, the reported rates were between \$4.00 and \$5.50, and in Boston, the Executive Officer of the Builders Association estimated that the going rate was less than \$6.00, compared to \$7.65 in the union sector. Some of these employers provided health insurance, paid holidays, and bonuses, but as in Erie County, it did not appear that the value of these benefits matched that represented by hourly contributions to union trust funds.

In two of the cities visited, Providence and Rochester, the Builders Association had recently conducted membership surveys

TABLE 46. *Homebuilding Industry*
Average Hourly Wage Rates of Builders in Selected Occupations
Rhode Island, December 1970

Occupation	Average Wage Rate Lowest	Highest
Finish carpenter	$3.81	$4.80
Rough carpenter	3.30	4.06
Laborer	2.61	3.12
Bulldozer operator	3.75	4.25

Source: Rhode Island Builders Association (NAHB).

Note: The number of respondents is 74.

in which average wages for various occupations were calculated. The surveys also provide some information on supplementary benefits in those cities. The Providence survey covered 74 builders out of 163 members of the Rhode Island Builders Association. The responses were weighted by size, and represent the going rates as of December 1970. Builders were asked to report their highest and lowest rates for four occupations: finish carpenters, rough carpenters, laborers, and bulldozer operators. The weighted averages are shown on Table 46. Union scales in the Providence area, as of July 1, 1971, were $6.775 for carpenters, $5.55 for carpenters on residential work, and $5.50 for laborers. There is no specific quotation for bulldozer operators, but the union scale for operating engineers on building construction ranged from $7.00 to $8.00. The survey also solicited information on various benefits provided by respondents. These are summarized in Table 47. By way of comparison, hourly fund contributions in the union sector totaled fifty cents for carpenters and forty-five cents for laborers.

The Rochester survey, however, conducted during 1971, covered many more occupations than the Providence survey. The tabulated figures do not indicate whether the averages are weighted, nor do they show the number of respondents on which the averages were based, but they do serve to depict the neighborhood of residential wage levels. Table 48 outlines the average wage calculated for each trade and, where available, the union scale and benefits. As for supplementary benefits, the survey concluded that only some tradesmen—carpenters, plumbers, and

TABLE 47. *Homebuilding Industry*
Benefits Provided by Builders
Rhode Island, December 1970

Benefits	Number of Builders	
	Yes	No
Paid holidays:		
New Year's Day	35	39
Memorial Day	22	52
Independence Day	25	49
V-J Day	21	53
Labor Day	27	47
Columbus Day	19	55
Armistice Day [a]	25	49
Thanksgiving [a]	37	37
Christmas	37	37
Paid vacations: [b]		
Finish carpenters	32	42
Rough carpenters	29	45
Laborers	22	52
Funeral leave	33	41
Sick leave	10	64
Jury duty	9	65
Travel time	19	55
Bonus		
Christmas	62	12
Other	11	63
Paid health insurance	25	49

Source: Rhode Island Builders Association (NAHB).

Note: The number of respondents is 74.

[a] Estimated from responses to previous year's survey.

[b] Most are for one week.

electricians—received health and welfare benefits, ranging from eleven to fifteen cents per hour. The only benefits reported were vacation time and truck or car allowance.

TABLE 48. *Homebuilding Industry Residential Building Average Hourly Wage Rates, Union Scales, and Benefits, by Occupation Monroe County, New York, July 1971*

Trade	Residential Rate (Nonunion)	Union Scale	Union Benefits (Total)
Skilled:			
Bricklayer	$5.75	$8.135	$0.80
Carpenter	4.50	7.76	0.905
Electrician	4.35	8.20	0.60
Machine operator [a]	4.00	6.265-7.47	1.10
Painter	4.33	7.31	0.69
Pipefitter	3.53	8.77	0.765
Plumber	4.30	8.77	0.765
Roofer	4.25	7.62	0.53
Sheet metal worker	4.03	8.08	0.835
Tile setter	4.00	8.74	—
Unskilled:			
Laborer	2.93	6.58	0.925
Mason tender	3.75	6.58	0.925
Tile setter's helper	3.12	7.65	0.15

Source: Rochester Builders Association (NAHB).

[a] Residential rate is for bulldozer operator. Union rate refers to operating engineers in building construction.

Other Comparisons

In the fall of 1971, the Labor Relations Department of the National Association of Home Builders asked its local affiliates to estimate the residential wage rates for various trades in their areas. The Association received responses from 71 Executive Officers, with 21 reporting only the local union scales. Most of these responses were from cities in which residential construction is heavily unionized. Unfortunately, most of the remaining responses came from smaller communities for which union rates are not published. Union-residential comparisons, based on this survey and Bureau of Labor Statistics data, are made for thirteen areas in Table 49. It must be stressed that the residential rates in most cases were not arrived at through formal surveys, but

TABLE 49. *Construction Industry*
Residential and Commercial Wage Rates and Benefits for Selected Trades
In Selected Cities, July 1, 1971

City	Carpenters			Bricklayers			Electricians		
	Resi-dential	Com-mercial	Benefits	Resi-dential	Com-mercial	Benefits	Resi-dential	Com-mercial	Benefits
Atlanta, Ga.	$4.50	$6.65	$0.35	$4.25	$6.45	$0.61	$4.75	$7.65	$0.60
Birmingham, Ala.	3.70	5.40	0.23	5.60	6.25	0.28	3.15	6.20	0.21
Boston, Mass.	6.25	7.65	0.60	6.25	8.10	0.66	6.25	7.85	1.23
Charlotte, N.C.	3.15	5.10	0.25	—	5.35	0.15	4.50	5.55	0.235
Chattanooga, Tenn.	4.75	5.73	0.32	5.00	6.25	0.50	5.75	6.55	0.265
Columbia, S.C.	2.50	4.30	—	3.00	n.a.	n.a.	3.50	5.10	0.35
Denver, Colo.	4.75	6.565	0.95	—	8.05	1.15	—	7.22	0.29
Fargo, N.D.	4.50	5.15	—	—	8.05	0.65	6.25	6.65	0.80
Grand Rapids, Mich.	5.50	6.85	0.77	6.00	6.70	0.45	6.00	7.37	0.37
Louisville, Ky.	3.75	7.10	0.47	4.50	7.58	0.69	4.25	8.08	0.28
Syracuse, N.Y.	3.75	7.71	0.935	—	7.99	0.91	—	8.40	1.105
Tampa, Fla.	4.25	5.78	0.325	4.00	6.20	0.48	4.25	6.50	0.215
Wichita, Kans.	4.50	6.75	—	5.50	6.73	0.85	—	7.30	0.07
Unweighted average	4.30	6.21	0.40	4.90	6.97	0.61	4.87	6.96	0.47

TABLE 49. *Construction Industry* (Continued)

City	Laborers			Painters			Plumbers		
	Resi-dential	Com-mercial	Benefits	Resi-dential	Com-mercial	Benefits	Resi-dential	Com-mercial	Benefits
Atlanta, Ga.	$2.50	$4.10	$0.20	$4.50	$6.85	$0.25	$6.00	$6.85	$0.57
Birmingham, Ala.	2.75	3.70	0.28	4.25	5.75	0.20	5.00	7.00	0.54
Boston, Mass.	4.25	6.10	0.50	4.50	6.96	0.67	6.50	8.20	0.69
Charlotte, N.C.	2.30	4.50	—	—	4.00	0.30	4.15	5.75	0.25
Chattanooga, Tenn.	3.50	4.00	0.20	4.70	5.45	0.10	6.00	7.10	0.40
Columbia, S.C.	2.00	n.a.	n.a.	3.25	4.00	0.30	3.50	5.90	0.20
Denver, Colo.	3.25	4.25	0.63	5.50	6.68	0.66	—	6.50	1.10
Fargo, N.D.	3.00	4.20	—	5.00	5.45	—	—	6.87	0.45
Grand Rapids, Mich.	3.50	5.21	0.70	5.75	6.10	0.45	5.75	7.64	0.57
Louisville, Ky.	2.50	5.45	0.47	4.00	5.74	0.25	5.50	6.98	1.255
Syracuse, N.Y.	2.88	6.70	1.025	3.50	6.80	0.75	—	7.85	0.88
Tampa, Fla.	2.50	4.175	0.15	3.50	4.85	0.15	4.50	6.70	0.65
Wichita, Kans.	3.00	4.90	0.60	3.75	5.45	0.10	5.00	7.61	0.87
Unweighted average	2.92	4.77	0.40	4.35	5.70	0.32	5.19	7.00	0.65

Source: U.S. Bureau of Labor Statistics, *Union Wages and Hours: Building Trades, 1971*, Bulletin 1747 (Washington, D.C.: Government Printing Office, 1972); "1971 N.A.H.B. Labor Survey" (unpublished).

rather through educated guesses of local industry spokesmen Nonetheless, the relative wage levels are roughly consistent with those found in Buffalo, Rochester, and Providence.

Table 49 also shows the unweighted averages of both union and residential rates in the thirteen cities covered. For mos trades, the union rate is at least 50 percent higher when fring benefits are included, although the addition of benefits to th residential rate would undoubtedly narrow the differentials Furthermore, if the union increases taking effect after Jul 1971 were to be added, the gaps would probably be even greater since recent market conditions in the general economy make i unlikely that nonunion rate increases have kept pace with thos in the union sector. It is also noteworthy that the differentia for laborers is substantially higher than those for the skille trades.

Thus far there has been no attempt to distinguish betwee residential wages and those paid in other segments of nonunio construction. It may well be that union-nonunion differentia in the commercial or industrial sector alone are different, bu unfortunately there are no sufficiently detailed data upon whic to base a meaningful judgment. Ongoing research by the In dustrial Research Unit, along with studies currently being con ducted by the Bureau of Labor Statistics should shed some ligh on the question. For the present, however, we must conten ourselves with an examination of nonunion residential rates relation to union scales.

Conclusion

Thus far we have attempted to assemble the available ev dence regarding the relative levels of union and nonunion wa rates, especially in the residential sector. Although this eviden is spotty and unsystematic, it is consistent in at least one maj respect: that nonunion scales generally, and residential scal in particular, are for the most part far below those paid in t union sector. Unfortunately, it is not possible to say definite how the differentials have behaved in recent years, although is worth noting that the studies discussed above suggest th the magnitude of the differentials in those years does not a pear to have changed appreciably. Still, two important questio remain: first, why these differentials develop and how th are maintained; and second, what their implications are the operation of labor markets in homebuilding. These questions are explored in the remainder of the chapter.

REASONS FOR UNION-RESIDENTIAL DIFFERENTIALS

As previously noted, the reason for the extraordinarily high wage levels found in union construction are highly complex and imperfectly understood. There is little doubt, however, that market power possessed and exercised by the building trades unions is an important element in the setting of wages. It is the judgment here that this power generates pay scales increasingly in excess of those "authorized" by market forces of supply and demand. Although some large settlements may in fact be consistent with tight labor market conditions in specific localities, the very spread of those settlements to areas in which construction demand is much less buoyant attests to the importance of "coercive comparisons"—a decidedly nonmarket phenomenon—in wage determination. The steady expansion of the differential between construction wages and those in most other industries (including some in which employment has been growing faster than in construction) suggests that somehow the extent of or the propensity to assert this power has grown. Unfortunately, as Mills observed, the source of any accretion of power is hard to identify, and the hypothesis that existing power has been used more aggressively over the years, assuming it is true, leaves open the question of why the power remained underutilized in the past.

We do not pretend to have the answers to these vexing questions. There are others, however, perhaps more germane to this inquiry. How, for instance, can the building trades unions retain and exert "monopoly" power in an industry one-half of which is unorganized? In part, the simple answer appears to be that they cannot, as evidenced by the growing concern of national union officials with the inroads of nonunion contractors. A second factor is that to some extent the *product* markets served by union and residential contractors are not entirely congruent. Homebuilding is an especially pertinent illustration of this point, since in many areas the unions have all but abandoned efforts to organize this sector. Since homebuilders, as noted in Chapter II, tend to limit their activity to residential construction, they do not pose a serious competitive threat to union standards in other branches of the industry.

To the extent that the markets of union and nonunion contractors do intersect, the maintenance of substantial wage differentials undoubtedly reflect some combination of both market and nonmarket factors. (A "market factor" refers to one which directly influences the supply and/or demand schedules for labor.) In a previous chapter, for example, it was argued that the prospect of steady work for regular construction workers—those who depend on construction work for their livelihood—is probably greater in homebuilding; thus, the uncertainty and irregularity of employment in the union sector necessitates a wage premium. In short, the security of a weekly pay check and the psychic benefit of not having to engage in job-seeking activity too often are partial substitutes for pecuniary compensation.

A number of respondents in the Erie County survey suggested that some workers regard the nature of the job and its environment as more interesting or rewarding in homebuilding. Most of these judgments came from builders rather than subcontractors, which is consistent with the observation frequently made by respondents that carpentry work on homes is generally more delicate and "craftsmanlike" than that on commercial structures. There were other perceived "rewarding" elements to house work. The following comments illustrate this.

—There is a lot more self-pride in workmanship in the homebuilding industry than in commercial construction. There is more personal contact between management and employees.

—They have more freedom in their work; they are not stuck on one phase of building. General feeling is that they like houses; moving from house to house is more interesting for them.

—Small builder is close to his employees—hunts, fishes, bowls together.

—Like the work better. It's more interesting. They feel personally involved in the construction. They like working with smaller groups of men.

—My workers would not go to the union because they are not carpenters —they just build forms. Here they're proud of their work and it's cleaner work.

—Personal contact has a lot to do with it. I'm always available; I'm one of the boys.

A few respondents suggested that workers resent union control over job opportunities and practices. Although these employers may have been merely projecting their own distaste for unions, it is certainly plausible that *some* workers prefer, for whatever reason, to work in a nonunion milieu and are willing to forego the higher wage. Thus the following comments were offered in response to the question: "Why do some workers continue to work in homebuilding rather than move into other higher branches of the industry?"

—Many do not like to work for unions or go to a business agent for a job. If you're in with the union and the right people you get the jobs; if you are not you don't get any work.

—They can't break into higher paying construction. Chosen few run each independent union and show favoritism to their friends and a lot of people get sick of this. If you don't know the right people you don't get along in these unions.

—Basically you don't have regimentation by unions and a lot of people resent this so they stay in homebuilding.

Finally, a handful of employers opined that their workers could not move into the union sector because they were not sufficiently qualified to perform the work. The argument suggests that higher wages in union construction result in a more competent and efficient work force, so that differences in total labor cost, if they exist at all, are less than the differences in hourly compensation. There is, in fact, some empirical support for this proposition. In 1957, Allan B. Mandelstamm compared labor costs in residential construction in two Michigan cities. At that time, Ann Arbor was strongly unionized and Bay City was predominantly nonunion except for one or two trades.[149] He asked contractors and subcontractors in both cities to estimate the man-hour requirements and the total wage and benefit bill for a specified task. The result was that the average number of man-hours cited by the Ann Arbor contractors was significantly lower than those in Bay City, so that despite higher wage rates in Ann Arbor, the average total wage bills were very similar. Furthermore, Mandelstamm found that labor requirements for nonunion contractors in Bay City were greater than those for union employers there, although the total wage bill for union contractors was still appreciably higher.[150]

[149] Allan B. Mandelstamm, "The Effects of Unions on Efficiency in the Residential Construction Industry: A Case Study," *Industrial and Labor Relations Review*, Vol. XVIII (July 1965), pp. 503-521.

[150] In this connection, the reader may be interested in a somewhat different kind of study in which it was concluded that higher wages in concentrated (and unionized) industries are largely attributable to "quality" differences in their work forces. Leonard Weiss, "Concentration and Labor

The Mandelstamm study suggests that some portion of union-nonunion differentials may indeed reflect differences in worker quality, although that is admittedly not the only factor. He found evidence of greater managerial efficiency among the Ann Arbor contractors, in part spurred on by the higher wages they had to pay and by the presence of an "efficient competitive leader" from the Detroit area. Furthermore, his findings on average labor costs and man-hour requirements worked out to a wage *rate* differential of only 23 percent, substantially lower than the range of differentials found in the studies cited earlier. Although the relationship between wages and efficiency demonstrated in that study provides at best a partial answer, it does serve to point up the need for a systematic attempt to develop a skill inventory in the various branches of construction.

The distribution of skills in the construction labor force is a question which has never been thoroughly investigated. Although some available data series do distinguish both broad occupational groups and individual crafts, the skill composition *within* crafts is not well-known, nor is the extent to which workers are capable of performing in more than one craft. Gross data on the *occupational* distribution of the work force are only marginally helpful. For example, although some carpenters are capable of only the roughest kinds of carpentry work, such as framing houses or building concrete forms, others are not only adept at delicate trim work but also can function as cement finishers, painters, bricklayers, or machine operators. Gross occupational figures, moreover, are not available for the homebuilding sector alone.

In interviews with industry spokesmen, their views were solicited on the relative skill requirements of given crafts in residential and nonresidential construction. Although skill requirements are not immutable—most employers are able to make do with marginally trained workers when highly skilled ones are scarce—one can still discuss the question in terms of the *range* of skills involved in different kinds of projects. The term "skill," moreover, is not without ambiguity; in different contexts and in different crafts it will encompass different qualities, such as strength, dexterity, knowledge, agility, endurance, or speed. Nevertheless, when the question was put in the broadest terms, the reactions were quite consistent. Except for carpentry, and

---

Earnings," *American Economic Review*, Vol. LVI (March 1966), pp. 96-117.

to a lesser extent masonry, most of the spokesmen regarded nonresidential work as more demanding; residential carpentry was seen, on the whole, as requiring more skill, although it must be recognized that even residential carpentry contains many tasks which can be learned in very short order by a worker with an average mechanical aptitude. Thus the *average* union-nonunion differential may be in part a result of greater skill requirements in those sectors where unions are stronger (e.g., commercial and industrial) than in those where most of the work is performed on a nonunion basis (e.g., residential).

To the extent that wage differentials reflect workers' preferences for residential employment or differences in their productive capacities, the phenomenon is a function of market forces. In other words, it would occur even in the absence of union monopoly power, as workers theoretically seek to maximize their "net advantage" in a job rather than simply the wage rate. The market explanation, however, is far from totally satisfying. About one-third of the Erie County respondents contended that their workers would move if they could; that is, if the union sector provided employment opportunities. It strains credulity, moreover, that the psychic advantages of residential work are such that large numbers of workers voluntarily forego a 50 to 100 percent increase in pay.

The classical explanation of a union-nonunion wage differential is that union power drives wages above market levels and thereby reduces the demand for labor, thus forcing workers into the nonunion sector which depresses wages there. The evidence relating to the exercise of that power by the building trades has already been discussed. Still, the differential can be maintained only if intersector mobility is limited. In construction, various contractual and informal procedures prevent employers from hiring outside the bargaining unit as long as members are out of work. Preferential hiring based on residency and previous work in the bargaining unit, along with contractors' reluctance to risk union retaliation (such as meticulous policing over provisions of the agreement or failing to dispatch qualified workers) for favoring outsiders over members are cases in point. Thus the differentials will not diminish until demand is high enough in the union sector to admit workers from homebuilding and other nonunion activity, until they become large enough to encourage nonunion contractors to enter the market, or until union contractors change to nonunion status. In this connection, the unions' ability to maintain their advantages will depend on their success in organizing the nonunion contractor. In short, although there are surely some market elements present, the differential is also a function of (1) the bargaining strength of the building trades unions (deriving from sources amply discussed in the literature); (2) their ability to exclude nonmembers from the job market when work opportunities are limited; and (3) the absence, until recently, of a successful effort by nonunion contractors, supported by purchasers of construction, to bid away what was once "union" work (related, no doubt, to the outsized union wage settlements of the 1968-1972 period). Unfortunately, the relative influence of the market and nonmarket factors cannot be readily quantified, although the admittedly unscientific impression here is that union power, at present, is the predominant element.

IMPACT OF UNION WAGES ON THE HOMEBUILDING INDUSTRY

Since most home construction is performed on a nonunion basis, the influence of collective bargaining on residential wages is, for the most part, indirect. In two respects, however, union wage setting presents a direct and immediate impact. The first, of course, is that in some areas homebuilders are themselves almost entirely unionized and pay wages identical to those in other sectors of the industry.[151] Given the very substantial differentials between housing and commercial construction that typically prevail whenever homebuilders are nonunion, the conclusion is inescapable that unionization of homebuilders has a dramatic effect on residential wages. Unfortunately, there is no way at present to ascertain with precision the impact of these collectively bargained wages on housing *costs,* although it seems highly unlikely that the kinds of differentials discussed earlier are balanced by productivity differences.[152]

Union wages may also impinge upon homebuilding through the operation of prevailing wage laws. The federal Davis-Bacon Act and similar state statutes require that wages and benefits

[151] Some agreements provide for lower rates on residential work, although the differences are far less than those between the union and nonunion sectors.

[152] For a *longitudinal* study of productivity improvements in residential construction, see Sara Behman, "On-Site Labor Productivity in Home Building," *Industrial Relations,* Vol. XI (October 1972), pp. 314-324.

on projects constructed with public funds must be at least equal to those "prevailing" in the area for each trade employed on the site. Under the federal law, the determination of the prevailing wage is made by the U.S. Department of Labor which sets each rate according to the following criteria: (1) that paid to the majority of workers in the area in each occupational group; (2) otherwise, that paid to the largest number of workers as long as that number is at least 30 percent of total employment in each group; and (3) otherwise, the mean rate paid in the area.

The Davis-Bacon Act was originally passed in 1931 to prevent competitive bidding on government projects from depressing wage standards and from encouraging the importation of workers from distant, lower-wage areas. Critics have contended, however, that its administration has not reinforced union rates where they are in fact "prevailing," but has also imposed union standards in areas where actual wages are appreciably lower.[153] This criticism has taken two major forms. First, it is argued that the areas for which the prevailing wages are calculated do not always conform to the labor market boundaries in which the project is constructed, with the result that union rates are in effect "exported" to entirely different markets;[154] and second, it appears that the Department of Labor has applied rates from one branch of construction to different types of projects. As the General Accounting Office concluded,

> In many cases, the Department has not distinguished between different types of construction, such as commercial and residential, although significant variances exist between labor rates applicable to these two types of construction. Often wage determinations have applied the higher rates for commercial-type building construction and have disregarded the rates for residential-type construction.[155]

---

[153] See especially D. M. Gujurathi, "The Economics of the Davis-Bacon Act," *Journal of Business*, Vol. XL (July 1967), pp. 303-316; John P. Gould, *Davis-Bacon Act: The Economics of Prevailing Wage Laws* (Washington, D.C.: American Enterprise Institute, 1971); and General Accounting Office, *Need for Improved Administration of the Davis-Bacon Act Noted Over a Decade of General Accounting Office Reviews* (Washington, D.C.: Government Printing Office, 1971). For a defense of prevailing wage laws, see David B. Johnson, "Prevailing Wage Law," in Gerald G. Somers (ed.), *Labor, Management, and Social Policy* (Madison: University of Wisconsin Press, 1963) pp. 231-262.

[154] Gujurathi found, for example, that 38 percent of determinations based on union rates (most of the total) came from statewide averages or *noncontiguous* counties. Gujurathi, *op. cit.*, p. 307.

[155] General Accounting Office, *op cit.*, p. 2.

These practices clearly seem to fly in the face of the criteria set forth in the Davis-Bacon Act itself, which requires the minimum wages to be those "prevailing for the corresponding classes of laborers and mechanics *employed on projects of a character similar to the contract work in the city, town, village or other subdivision of the State in which the work is to be performed.*"[156] It is true that some projects are of such a character that there are no comparable ones in the immediate area, and it may also be argued that in some areas "the negotiated rate in effect is the only rate for which labor can be obtained in quantities necessary to perform the work. As such, it has to be the prevailing rate."[157] But these observations would not seem to apply to public housing projects in metropolitan areas where most residential construction is nonunion.

Even where prevailing rates are set below union scales, they may still be substantially higher than those actually paid in an area. We alluded earlier to a residential wage survey conducted in Rochester by the Home Builders Association. That survey was made in response to a series of prevailing wage determinations established by the New York State Bureau of Public Works for a housing project in a Rochester suburb. The established rates, although somewhat lower than the union standards, were much higher than the survey indicated was the practice. For example, residential rates for carpenters were found to average $4.50, compared to the determined rate of $7.66 and the union rate (including fringe benefits) of $9.41. The ratios were roughly the same for most of the other skilled trades.

At the same time, it is doubtful that prevailing wage laws have as much impact on average residential wages or on housing costs as these criticisms might suggest. In the first place, public housing is only a small fraction of the aggregate. In 1972, for example, expenditures on public housing and redevelopment construction totalled $1.9 billion out of $55 billion spent on all residential buildings.[158] Furthermore, as Mills has noted, "Labor costs as a whole may be minimized under a Davis-Bacon determination by the employers assigning work to employees in lower wage classifications than is common practice in

---

[156] *Ibid.*, p. 13.

[157] Johnson, *op. cit.*, p. 249.

[158] U.S. Department of Commerce, *Construction Review*, Vol. XIX, no. 10 (October 1973), pp. 7-8.

the industry."[159] Indeed, it is possible that nonunion employers may enjoy cost advantages even while paying union rates, owing to their greater flexibility in allocating the work. Although it is difficult to escape the conclusion that prevailing wage laws have to some degree raised average residential wages and labor costs, either by discouraging nonunion contractors from bidding on public housing jobs or by forcing them to pay more than their normal rates on such jobs,[160] their limited applicability in the housing sector suggests that their direct impact has not been major. The influence of prevailing wage laws on *nonresidential* nonunion construction is probably much greater. The Industrial Research Unit is presently conducting a comprehensive study of the Davis-Bacon Act and related statutes which should allow more definitive conclusions on this subject.

CONCLUDING REMARKS

It is extremely difficult to forward any conclusions as to the indirect effects of collective bargaining on residential wages. The question is whether homebuilding rates are higher than they would be in the absence of bargaining in other sectors of the industry. With detailed longitudinal data, the researcher could analyze the movements of residential rates in response to union settlements and in the context of relative market conditions in various industrial classifications. Unfortunately, as we have seen, these data do not exist.

An alternative approach would be to examine union and nonunion wages on a cross-sectional basis. Other things equal, the union influence on nonunion wages should be reflected in smaller differentials where unions are stronger. In other words, we would expect that the union influence would "spill over" into the nonunion sector where the unions have greater control over the construction labor market. There are, in fact, a few small clues that this is the case. The Bureau of Labor Statistics study of annual earnings in 1970, for example, found larger differentials for laborers (who are less completely organized than

---

[159] Mills, *op. cit.*, p. 82, fn. 45. It should be noted, however, that such action may violate Davis-Bacon regulations.

[160] Of course, the possibility cannot be ignored that some nonunion contractors will pay the determined rate on *all* their jobs, in order to obtain the government contract and at the same time avoid the personnel problems attendant to paying different wages for similar work. Unfortunately, there is no evidence as to the extent of this practice.

craftsmen) than for craftsmen, and larger differentials in the South (where unions are relatively weak) than in the rest of the country.[161] Of course, however, other things are not necessarily equal, and it would be hazardous to draw any firm conclusions from such exiguous data.

It would also be helpful to have union and nonunion rates for the various industrial divisions of construction. Are nonunion rates of highway or large building contractors, who are in more direct competition with union employers, closer to the union scale than those of homebuilders? Most persons knowledgeable about the industry would probably agree that they are, but again there is no comprehensive evidence. In short, there is no way at present to make an empirical case that unions help drive up nonunion wages, although elementary labor market analysis suggests that, in theory, there will be a spillover effect under certain product market conditions.

Finally, it must be emphasized that whatever indirect effects the unions may have, those effects will not necessarily be spread throughout the residential work force. The reader will recall that residential wage structures are quite dispersed, even within a specific occupational category. The wage data used throughout most of this chapter are *averages.* The effect of union wages may well be to pull up the pay scales of only a fraction of all nonunion workers, those who would be most able to command the rates paid in union construction. In other words, that very dispersion of nonunion rates may be in part a function of union-nonunion competition for the better workers. The ability of the nonunion contractor to fix his rates to the individual rather than to the group, therefore, cushions the impact of union scales on his total wage bill. The relatively high differentials for unskilled labor (even where such labor is dominantly unionized in commercial construction, as in the Buffalo area) are further testimony to this point, for in the absence of a generally tight labor market throughout the area, the nonunion employer need not fear running short of untrained workers. Thus, even to the extent that unions influence wages in homebuilding, the builder is partially insulated from union competition by the very fact that he does not have to pay a uniform rate.

---

161 It will be recalled that this study referred to year-round workers, so annual earnings are reasonably reflective of wage rates. U.S. Bureau of Labor Statistics, *Selected Earnings, op. cit.,* Tables 6, 10.

CHAPTER VII

Conclusion

As acknowledged earlier, this study has barely begun to scratch the surface of labor market practices in nonunion residential construction, and as such provides only a most rudimentary understanding of how these markets function. The scholarly research on construction labor has dealt with the industry as though the nonunion sector were a minor appendage. This assumption, as is becoming increasingly manifest, is untenable, and accordingly the Industrial Research Unit has initiated a major research effort to delve a good deal deeper than was possible here into the structure, behavior, and impact of nonunion construction in both the residential and the nonresidential sectors.

This final chapter will constitute a bridge between the exploratory study discussed in the foregoing pages and the more comprehensive studies to follow. The object here is to highlight the salient findings thus far, and to identify the larger areas still to be investigated. As we shall see, although a survey of one industry segment in a single metropolitan area does not permit definite conclusions about nonunion construction, it does help identify and define research issues to be examined on a broader scale.

The economics of the labor market, broadly construed, enable human resources to be recruited, developed, and allocated in response to sets of factors which determine the need for manpower and the propensity of workers to offer their services. In the organized sector, these processes are greatly influenced by the political exigencies, real or imagined, faced by the unions and their responses to them. In nonunion construction, they are determined by the economic decisions of individual employers and workers. The ways in which and the constraints under which these decisions are made have constituted the central concern of this volume. The issues considered here have not exhausted all the important elements of labor market operations.

Some of the omissions were acknowledged in the opening chapter; others will be noted below. Still, a beginning has been made.

In studying labor markets and labor relations in construction—even when limited to the nonunion sector or a portion of it—the analyst must consider carefully the heterogeneity of the industry. Its diverse constituent parts must be recognized in exploring virtually any aspect of the industrial relations system. It cannot be assumed, for example, that the behavior and experience of the homebuilders' labor market is similar to that of industrial or commerical contractors'; that the nonunion firm which builds restaurants, service stations, or garden apartments is similar to the firm which erects twenty-story office buildings; or that the practices of Ace Construction Company of Anytown, U.S.A., are also followed by Brown and Root. Correspondingly, labor market parameters are not necessarily the same for different specialties: the masonry contractor, for example, may well have one type of relationship with both his employees and the general contractor but different from the electrical contractor. The differences by size and type of employer impinge upon all the issues to be outlined below.

As a general proposition, the various elements of an employer's labor market behavior can be viewed as the methods by which he responds to some change in his manpower requirements. The change may relate to the size of the needed work force or its composition. It may be seen as temporary, when seasonal fluctuations require additional workers for a few months, or relatively permanent, when a full-time worker leaves and must be replaced or when the overall level of demand for the employer's product grows. The response may bring the employer in contact with the external labor market, or it may take place internally, through a reconstitution of jobs, a reallocation of assignments, and/or a reclassification (upgrading) of particular workers. Both the change and the response will be related to the matter of cost; it may be expected that labor market operations are different under one set of wage parameters from those under another.

The results of this survey have suggested that many labor market functions in residential construction are marked by their informality and flexibility. This finding should occasion little surprise, for much previous writing on construction labor has remarked on the volatility of the industry's needs. Where unions are influential, however, the employer's behavior is in some respects constrained; in the nonunion sector, on the other hand, the responses to change appear to be different in degree, if not in kind.

LABOR MARKET PRACTICES: RECRUITMENT AND ALLOCATION

The recruitment of workers by residential employers in Erie County relies heavily on an informal network of communication among employees, contractors, subcontractors, and suppliers. This use of the "grapevine" was also cited as the dominant element by spokesmen in the other localities visited as part of the study. There does not appear to be any central source for job information which serves to summarize information on the state of the market on either the supply or demand sides. In some places, local homebuilders' associations occasionally attempt to match workers and jobs, but in no locale was the association widely recognized and used as a clearinghouse of information. The utilization of formal employment agencies, either public or private, was infrequent. The most systematic avenue of search used to any appreciable extent is newspaper advertising, and even this method tends to be more of a factor in the recruitment of unskilled labor than qualified craftsmen.

The question of recruitment methods and their effectiveness is particularly important in the construction labor market, where turnover is high, and needs unstable. Simply put, there is a substantial amount of hiring, and the smoothness with which the matching takes place—or, stated otherwise, the time it takes—is an important determinant of the efficiency of the market mechanism. In addition, the matching process is hampered by the dispersion and abundance of firms in the industry. Under these conditions, one may well question whether the "grapevine" in any one market can handle the filling of large numbers of openings with optimal expedition.

Employers in Erie County were generally receptive to the idea of a central employment exchange operated through the local builders' association, although the notion was not without its critics. The principal reservation related to the effect of such an arrangement on the prospect of unionization. Some respondents felt—and the feeling was echoed elsewhere—that any move toward greater centralization of the industry might invite union incursions. It is not immediately clear, however, how a straightforward up-to-date record of job openings and available workers which is accessible for voluntary consultation by both workers and employers would make union organization easier. This apprehension clearly exists and would have to be explicitly addressed before any such information exchange could be instituted.

Differences in recruitment methods among the various types of employers in the homebuilding industry are areas that sorely need further exploration. The sample for this study was too small to permit fine distinctions. It would be especially interesting to learn the methods of out-of-town contractors who do not rely so completely on the grapevine. Similarly, it is likely that employers of certain trades (e.g., electricians and sheet metal workers) approach the problem differently than do those of trades requiring a somewhat lower order of skill (e.g., carpenters and painters).

This study has not explored in detail the question of work allocation. The ability of the nonunion contractor to deploy his work force with maximum efficiency has been considered an important factor in the growth of nonunion construction along with the differences in wage rates. This issue really has two components: (1) to what extent can unskilled labor be utilized to perform certain portions of the work normally associated with a craftsman? and (2) to what extent are journeymen sufficiently adept at more than one trade to reduce the employer's reliance upon the external market when there are temporary labor shortages? In short, an analysis of the internal labor market is a vital complement to one of hiring procedures, and such an analysis must reach beyond formal occupational classifications, such as "carpenter" or "laborer." The nature of the internal market mechanism is largely dependent upon the inventory of skills available to the employer rather than simply the numbers and titles of tradesmen in his employ.

Both the internal and external markets are involved in the question of seasonality. The results of this study have suggested that seasonal patterns in employment and unemployment are somewhat different in the residential sector than in the more strongly unionized portions of the industry. The difference devolves largely from the greater flexibility enjoyed by the nonunion contractor in supplementing his regular work force with temporary employees and in redeploying his workers so as to derive maximum utility from those with marginal skills. Thus it seems likely that the peak-season work force of the residential employer is comprised to a greater extent of students and other part-year workers, with his regular employees subject to fewer and shorter spells of unemployment in the off-season. Again, these patterns will vary with the type of contractor and the experience of some nonresidential contractors who operate on a nonunion basis may be somewhat different, especially if the firm is functioning away from its home base.

Erie County employers have provided some important indications as to how seasonal changes in demand are met, but definitive conclusions must await a broader inquiry. For example, the sources and skill levels of temporary workers are not completely understood, nor is the extent to which these temporary workers ultimately join the labor force on a permanent basis. In other words, it may be that the opportunities afforded to untrained entrants by seasonal needs is an important element in the development of skills, as temporary employees learn a trade gradually by repeated exposure to construction work over several summers. It is also important to learn the extent to which other adjustments, ones which are less economic or feasible for the union employer, are utilized. Under union contracts, for example, work by supervisors may be prohibited and extensive overtime prohibitive. This study suggests that in homebuilding such practices are rather common. Finally, there are the adjustments on the other end of the seasonal spectrum. If layoffs are generally less severe in the nonunion sector, the implications are important for all of the industry, since intermittency of employment represents a major waste of resources. We need to know more about the methods by which nonunion employers actively attempt to stabilize employment, if indeed they do.

Still another labor market issue—one which has been almost totally neglected here—is the nature of subcontracting. Residential builders in Erie County appear to contract out a larger proportion of their work on an average than do other general contractors. It is not clear whether this phenomenon relates more to the technology of building homes or to the nonunion status of the market. In either case, the relationship between a contractor and subcontractor may take a variety of forms. Some subcontractors are little more than employees of the builder operating under an incentive system (e.g., piecework); others are entirely independent firms. Subsequent research should attempt to elucidate the various types of contracting relationships and to learn the extent to which practices differ among specialties, geographic areas, and firms of different size and industrial market. It may be important to learn whether some kinds of contracting relationships are entered into in order to maximize efficiency or rather to hold down wage costs.

LABOR MARKET PRACTICES: SKILL DEVELOPMENT

If the findings of this local study are representative of the situation nationally, it follows that formal training practices are few and far between in residential construction. Whether Erie County reflects conditions throughout the country, or whether the residential sector reflects nonunion operations generally, are open questions. Local efforts to develop formal training programs in the cities visited were, at best, small-scale and ephemeral. That this situation is representative of nationwide practice is strongly suggested by the membership survey of the National Association of Home Builders. Moreover, formal training programs sponsored by the NAHB have reached only a fraction of the nonunion residential labor markets. However, it appears from the Erie County study that individual employers do not provide or encourage the use of structured training devices to generate skills.

In short, the conclusion seems inescapable that the proportion of residential craftsmen who have undergone formal training of any appreciable magnitude is extremely small. By and large, skills are acquired gradually through the production process itself, as the untrained laborer or helper learns a trade by observing the work of others. Although it cannot be denied that such learning may well produce an able journeyman, it is likely that the skills developed will be narrow and specialized and that the worker's ability to function without close supervision limited. Furthermore, it may be argued that the learning process itself is slower in the absence of a logical sequence of tasks to be mastered, and the lack of a known and steady stream of adequately trained workers may hamper the industry's ability to respond expeditiously to rapidly changing needs.

This study has identified various training avenues outside residential construction, but it has left unanswered the ways in which these methods relate to skill development within the industry. It is not clear, for example, how well vocational education prepares young persons for subsequent learning in job-related environments, or whether skills acquired in other industries are readily transferable to homebuilding. In any case, however, the major contributions to skill development are probably made by the industry itself, either through the few formal training programs that have been initiated or through the undirected and unstructured acquisition of skill gained on the job.

The incidence of formal training in nonunion construction is gradually developing. Training is conducted not only under the auspices of the major employer associations, but also by some individual contractors. Recently, nonunion apprenticeship programs have registered with federal or state apprenticeship agencies, and thus the picture of the past may change rapidly. At present, we know very little about the administration of these programs, what crafts are available for training, what methods are used for selecting trainees, and—perhaps most importantly—what the trainees find once they have completed the program.

Apart from the industry-sponsored programs, which despite their recent growth still undoubtedly reach only a tiny fraction of new entrants, we need to know much more about the training activities of individual employers. It is doubtful that the Erie County homebuilding industry is representative of nonunion construction generally. One might reasonably expect, for example, that investment in training varies with the size of the firm. Since residential employers tend to be smaller on the average than those in other industry sectors, it is likely that nonresidential, nonunion contractors—especially the larger ones—make a greater effort to develop their own skilled labor. The extent and nature of this training will be covered in subsequent reports.

Most important, however, we need to know a good deal more about the informal training and upgrading mechanisms on which the homebuilding industry—and very likely the nonunion sector generally—depends heavily for the generation of the skills it needs. According to Erie County employers, the overwhelming majority of workers are trained on the job. It is very possible, however, that on-the-job training means different things to different people. A realistic assessment of skill development in the industry requires a more precise understanding of the nature of the learning process. For example, to what extent does this form of training in practice entail foregone production of journeymen who supervise and instruct the trainee? To what extent does it include manipulative instruction outside working hours? Or, alternatively, does it involve merely unguided observation by the learner, in which case the term "training" itself is a misnomer since there is no trainer?

Finally, we would hope to learn more about how internal upgrading procedures are affected by the state of the labor market. What determines the rate at which workers are given an opportunity to practice at ever more demanding tasks and what determines a worker's eligibility for promotion? It would also be useful to know whether most firms have a clearly delineated progression of occupational classifications—and wage rates—which reflect upward movement in the process of skill acquisition. Since the meaning of such designations undoubtedly varies among firms, they would not necessarily allow dissection and analysis of learning progressions. They would, however, cast some light on how employers view the learning function, in that a consistently applied series of job categories should reflect an expectation on the part of the employer that skill development is an integral part of the production process itself. In this context, it is important to know the extent to which upgrading within the firm is systematic or haphazard; whether movement up the job ladder is determined by a consistent set of performance criteria (and what those criteria are); and if progressive levels of attainment are explicitly recognized in job titles or wage rates.

EMPLOYEE COMPENSATION

The results of this study point strongly to two conclusions about wage rates in residential construction: first, that wages are on average substantially lower than union rates in the same labor markets; and second, that there is a wide range of pay scales to be found for any given trade within a market area. The data gathered in Erie County, moreover, suggest that even at the upper end of this range, wages are lower than those paid uniformly in the union sector. Although there were variations in the size of the differential, this finding held for all trades. By and large, it appears that the differential is inversely related to the skill requirements of the trade with the most dramatic differences to be found in the unskilled ranks.

This study has also revealed that supplementary benefits in homebuilding are both less extensive and less generous than those in union construction. In none of the localities visited were there any industry-wide benefit plans in operation. Some individual employers provide their workers with health insurance, paid holidays and vacations, and bonuses, but such practices are by no means universal. To some extent, these differences in total compensation may be balanced by greater opportunities for overtime and steadier employment over the year, but on the whole, their magnitude discourages the conclusion that even annual earnings are comparable.

The fragmented information available on wage rates in residential construction throughout the country suggests also that the Erie County situation is not extraordinary. At the same time, however, it does not necessarily follow that homebuilding is representative of the nonunion sector generally. Although there are some government data which indicate that average wages in union construction are substantially higher than the average paid by nonunion firms, these data do not take account of differences in labor force composition by trade or differences in skill levels within a trade. It is possible that the more proficient nonunion workers earn as much—or even more—than their union counterparts. In short, the averages may well conceal significant variations in the distribution of wage rates in a given market.

The more comprehensive study of nonunion construction forthcoming from the Industrial Research Unit will cast additional light on this issue. It will explore not only the range of pay scales in the industry, but also the patterns among trades and among different types of firms. Equally important, it will address the increasingly significant matter of wage supplements, both those offered by individual employers and those administered on a multi-employer basis.

The implications of wage levels and structures in nonunion construction go far beyond the academic interest of the researcher. To be sure, wage rates are only one element in overall labor costs and an even smaller factor in total construction costs. The extent to which nonunion wages, even after allowing for possible skill differences, however, are indeed lower than union rates is of great significance to both union and nonunion contractors, as well as purchasers of the construction product. It also has moment for policy makers, for it relates directly to the question of the extent to which union rates are consistent with market pressures and, alternatively, the degree to which they are the product of institutional power. Any economically logical stabilization program must obviously take cognizance of the relative weights of these factors.

There is another policy question involved as well in the wage structure of the industry. As noted in an earlier chapter, there has long been controversy over the administration of the federal Davis-Bacon Act and comparable state agencies. Strikingly, most of the debate has taken place without any real knowledge of actual wages paid in the nonunion sector. Thus the assembling of such data should allow a more competent assessment of the consistency—or lack of it—of Davis-Bacon administrative decisions with the spirit and standards set down in the statute itself. Furthermore, it will be useful to have a basis for evaluating the impact of Davis-Bacon determinations on the wage structure itself because an evaluation is difficult without more complete information on nonunion wages than is currently available. Finally, the larger study will explore the effect of the Davis-Bacon Act on the occupational structure of the industry, especially as it relates to the role of the "helper" in the production process.

MINORITIES IN CONSTRUCTION

This study has had very little to say about the pressing and controversial issue of racial discrimination in the construction industry generally, or homebuilding specifically. Since the field work did not expressly undertake to explore this matter, we are not in a position here to advance explicit conclusions concerning the relative participation of blacks and other minorities in residential construction. This section, therefore, is limited to a discussion of the implications of the foregoing labor market analysis on the prospects for encouraging greater minority participation, along with an outline of some of the more salient questions to be examined in subsequent research.

Many of the features of residential labor markets discussed in previous chapters relate to providing job opportunities for minorities. Some of these characteristics should serve to facilitate entry, whereas others may act to make it more difficult. In this context, we are primarily concerned with factors which distinguish residential markets from those in other branches of construction. Initially, however, it is worthwhile to outline certain barriers to minority employment which appear to be operative throughout the industry.

Many of these barriers stem to some degree from the cumulative effects of underrepresentation over the years. An individual's occupational choice is in part a function of the jobs held by others in his environment. Where minorities do not participate in a particular labor market, there is no model for individual members—especially the young—to follow in pursuing a career. Writing in the pre-Civil Rights Act era, Strauss and Ingerman made this point convincingly in regard to formal apprenticeship programs, but it is no less germane to more direct routes to the job. Specifically, the authors noted that "(1) qualified Negroes rarely perceive the skilled trades as a potential occupation; [and] (2) Negroes have little information as to how to apply for apprenticeship occupations."[162] Even to the extent to which negative perceptions and lack of information are not attributable to outright discrimination, they served to discourage blacks from seeking out and applying for jobs in construction. If overt discrimination is limited to the union sector, it is questionable whether the prospective applicant observing few skilled craftsmen of his race working anywhere in cosntruction will make the distinction. Thus, to the extent that access to labor market information as well as a perception of construction as a realistic job option has depended upon being plugged into the grapevine of the construction community, minority employment has been retarded, and as a result, alternatives to the grapevine method will be vital elements to any remedial policies.

Other factors operate to discourage minorities once they are on the job. One factor is the issue of social acceptance. Although there is no apparent reason to presume that social ostracism occurs more readily in construction than in other industries, the new worker who aspires to a skilled trade is particularly dependent upon his co-workers for help and guidance.[163] Furthermore, there is the instability and uncertainty of employment ever present to construction workers. These conditions make access to labor market information vital, access

---

[162] George Strauss and Sidney Ingerman, "Public Policy and Discrimination in Apprenticeship," *Hastings Law Journal,* Vol. XVI (February 1965), pp. 285-331.

[163] Strauss and Ingerman make this point with respect to the apprentice; it is equally true of the worker who relies on informal training.

which, as we have seen, may be a particular problem for the minority worker.

Factors Favoring Minority Employment

The considerations discussed in the foregoing paragraph do not necessarily have special saliency for residential construction; however, those outlined in previous chapters do. The relative flexibility enjoyed by nonunion builders in various labor market activities should serve to facilitate minority employment. Wage rates, for example, can be pegged to the capabilities of the individual worker, enabling blacks with narrow or marginal skills to be hired at rates commensurate with their productivity. The wide range of possible wages permits a gradual upgrading of new entrants without imposing a cost burden on the employer. In this context, the absence of prevailing wage requirements in most residential building has a similar effect. Thus, to the extent that underrepresentation of blacks is attributable to lack of sufficient skills, homebuilding appears to offer better employment prospects for blacks than those sectors in which uniform wages and fewer implicit job classifications are the rule.

Similarly, the homebuilder's latitude in allocating his work force without the impediment of strict occupational lines permits more flexibility in hiring. If workers can be readily shifted among various functions, there is also a greater likelihood of their being retained for longer durations. In the absence of craft barriers, it is more feasible for an employer to take on untrained or partially trained minority workers and assign them to a variety of tasks which, in the union sector, might cut across craft lines and engender jurisdictional resistance. In short, the greater flexibility both in setting wages and assigning work generates more of what may be termed "entry-level" jobs, into which unskilled or semiskilled workers can be easily absorbed, than does other building construction.

The relative stability of residential employment for many workers should also serve to ease minority entry which results in a more or less permanent attachment to the industry and, hence, some degree of employment security. This point is especially important in light of the frequent observation that blacks are underrepresented not in the construction work force as a whole, but only in the highly skilled trades. In many crafts, journeymen customarily learn their skills on the job, and the unskilled worker who is able to find only six months of employment a year may well find the learning process seriously impeded by lengthy interruptions, a barrier which both retards the upgrading of workers to skilled positions and discourages them from remaining in the industry. In short, the factors which favor the access of minorities to entry-level jobs may be abetted by relatively stable employment patterns which facilitate their advancement.

Another factor favorable to black employment in residential construction is the occupational mix. Carpenters, and primarily rough, framing carpenters, are by far the most numerous occupational group, followed by laborers, cement finishers, and sheetrock installers. These are all jobs in which blacks have historically found employment in contrast to the electrical, mechanical, and pipe trades in which discrimination has been the most overt and intense.[164] Moreover, even in these latter trades, the absence of unions which, at least until recently, have been traditionally inhospitable to the training and employment of blacks, increase opportunities for minorities in the overwhelmingly nonunion housing sector.

Finally, and perhaps most obvious, blacks are less likely to encounter overt discrimination in homebuilding. Although there is no reason to believe that nonunion white workers are any more racially tolerant or any less sensitive to the security of their own jobs than their counterparts in the union sector, the absence of unions removes a significant instrument through which exclusionary pressures can be exerted. Without these pressures, employers are more apt to base their hiring decisions strictly upon economic considerations. (On the other hand, these criteria may in some circumstances restrict black employment prospects; more than one observer has noted that minority contractors seldom make special efforts to employ nonwhite workers.)[165] Again, however, it should be noted that outright discrim-

---

164 See for background, Herbert R. Northrup, *Organized Labor and the Negro* (New York: Kraus Reprint Co., 1971), Chapter II.

165 See Richard L. Rowan and Lester Rubin, *Opening the Skilled Construction Trades to Blacks: A Study of the Washington and Indianapolis Plans for Minority Employment*, Labor Relations & Public Policy, Report No. 7 (Philadelphia: Industrial Research Unit, The Wharton School, University of Pennsylvania, 1972), p. 96; and Daniel Quinn Mills, *Industrial Relations and Manpower in Construction* (Cambridge: MIT Press, 1972), p. 172.

ination against *qualified* black applicants is now, on the whole, a less significant cause of their small numbers in the skilled trades than is their lack of training (which may be attributable to discrimination), and thus any advantages homebuilding may offer may well derive more from its labor market than from its racial neutrality.

Factors Unfavorable to Minority Employment

There are several characteristics of the homebuilding industry, relating in large measure to its structure and location, which would appear to militate against minority entry. In the first place, in most metropolitan areas new housing construction is heavily concentrated in the suburbs or country, whereas most potential black workers live in the central cities. This imbalance, although not insuperable, raises the cost of both job search and commuting to and from work. Even though it may be argued that a regularly employed worker should be able to afford an automobile, it must be remembered that much job information in nonunion construction is obtained informally, and access to this information may be expected to decline as the distance between home and work increases. (The significance of this factor, however, is not at present known. It would be helpful to learn, for example, whether minority representation is lower in an area where most nonunion residential construction is very far from the central city—such as Boston—than in an area where there is still considerable homebuilding in contiguous suburbs—such as Buffalo.)

As noted in Chapter II, building firms are highly dispersed in terms of industrial structure as well as geography. Although this is true of construction generally, it is particularly characteristic of the residential sector. Employers tend to be small and numerous. This dispersion may have a number of unfavorable consequences for minority hiring. In the absence of a central employment exchange, locating job opportunities may be expected to be more difficult than where a few employers in the area do most of the hiring. Furthermore, small employers are less able to absorb the costs of training and and upgrading; therefore, new hires may be relegated to repetitive, unskilled jobs with little opportunity for advancement. A firm with only one or two highly skilled mechanics is not in a good position to devote a substantial portion of its time to detailed supervision and instruction of untrained workers. Finally, it is arguable that the small employer with close social ties to his workers is more apt to engage in discriminatory hiring practices. As Mills suggests,

> It may well be that racial discrimination is most likely to occur in the small firm and in the presence of relatively stable employment relationships . . . . Larger contractors . . . given the impersonality of [their] employment relationships, are probably less likely to practice racial discrimination in employment.[166]

Mills, however, offers no basis for this observation. Small nonunion construction employers in the Philadelphia area, for example, were found to utilize a high percentage of blacks in a variety of skills. In fact, several advised Industrial Research Unit investigators that they depended heavily on blacks as permanent employees, augmented by students in the summer.[167] The major study of open shop construction, scheduled for release in 1975, will no doubt shed more light on this question.

The structural dispersion of residential construction, as well as its predominantly nonunion character, may also have served to limit the efforts to increase minority employment in another sense. As Rowan and Rubin found in both Washington and (to a lesser extent) Indianapolis, boosting minority employment in nonunion construction—especially in the skilled trades—has not been regarded with the same sense of urgency as similar objectives in the union sector. Nonunion contractors have been excluded for the most part in the development of local affirmative action plans.[168] Residential construction not only does not project an "image" of a bastion of exclusionary practices and, hence, has not had as much attention focused upon it as other sectors; but in addition, the federal government has not tried to, or has not been successful in, bringing the nonunion sector within the scope of such plans. As a result, it has not been subject to the same kind of pressures from governmental and private agencies to increase minority employment. (To be sure, the meagerness of publicly supported residential construction removes an important source of pressure in this regard.) At the same time, even if these pressures were to be exerted, the atomistic nature of the industry may well make the task of developing and monitoring hiring standards more difficult in many localities.

---

166 Mills, *op. cit.*, p. 153.

167 Based on field notes for forthcoming open shop construction study.

168 Rowan and Rubin, *op. cit.*, pp. 92-96, 162-164.

Some Preliminary Observations

The speculative nature of much of the foregoing discussion suggests what is perhaps the most important need, namely, a basic understanding of the level and patterns of minority employment in homebuilding. At the moment, we simply do not know enough about these parameters to devise detailed strategies for promoting opportunities for nonwhites in residential construction. For the present, however, it seems reasonable to presume that minorities are underrepresented among the skilled trades in homebuilding because they are underrepresented among higher status occupations in most other industries. With this presumption, and pending more systematic inquiry, some general comments can be made about strategies and approaches suggested by the labor market analyses in preceding chapters.

First, a critical prerequisite to participation in any labor market is access to information: what jobs are available; where are they located; what do they pay; and what are their skill requirements? This access, as we have seen, is especially important in homebuilding, where the transmission of information is largely carried out through an informal network of contacts and personal relationships. It is the judgment here that any organization with an objective to increase minority employment must act, as one of its principal functions, as a broker of job market information. It must serve as a nexus between the supply and demand sides of the market, performing intensive outreach activities to both potential workers in the minority community and prospective employers. In short, it is not enough to identify job openings and to find or develop workers with the necessary skills; the two must be brought together in some systematic fashion.

At the same time, of course, there are tasks to be performed on the demand and supply sides independently. Employers in the residential sector are not as well-organized as those in other parts of the industry. Furthermore, associations of nonunion contractors do not customarily spend as much of their time and resources on labor market matters. In Erie County, and undoubtedly elsewhere, many residential subcontractors are not even affiliated with any industry group. Thus enlisting the active support of local associations in promoting minority employment should be only a first step in the process. It will also be necessary in many communities to identify and solicit the cooperation of employers—primarily in the subtrades—who have not otherwise been deeply involved in industry-wide activities.

Given the relatively small involvement of residential contractors in government construction, it is unlikely that improvement in minority utilization can be accomplished through legal or economic pressure alone (although "moral suasion" may be a promising alternative). Instead, participation in such programs will probably be more effectively encouraged if suggested in terms of the industry's own self-interest. Affirmative action programs can be presented as a means of improving the efficiency of the residential labor market, by providing a reliable source of needed workers. Individual employers, in short, would be encouraged to view their participation in terms of a *quid pro quo* rather than as a one-way service they are being pressured or persuaded to offer.

On the whole, however, the evidence suggests that demand side considerations may not be paramount in residential construction. The relative stability of employment, the flexibility in wage setting and work force allocation, and the absence of organized resistance from unions serve to render the residential sector more open to minority entry. The feeling here is that the principal thrust in this area must be on the supply side. It is particularly vital that homebuilding be perceived by the minority community as an important source of sufficiently steady, well-paying jobs. It is difficult to make this point without sounding platitudinous, but as universities and other institutions are beginning to learn, affirmative action is an empty phrase without a ready supply of applicants toward whom the action can be directed.

Public agencies have not been inactive in developing and encouraging programs to prepare minority youths for careers in homebuilding. The federally supported training effort of the National Association of Home Builders is a case in point. The government should continue to assist the industry in developing an ongoing system for providing an adequate cadre of thoroughly trained mechanics. As we have seen, however, this avenue of entry is not numerically dominant. The bulk of the residential work force will in all likelihood continue to be trained on an informal and unstructured basis. In this connection, upgrading can take place only if the worker with an aptitude for the work and potential for advancement is induced to enter the work force. Such inducement, however, is difficult if entry-level jobs are viewed as a dead end by those prospective applicants.

Whatever the specific approach taken, efforts to enhance minority representation in the skilled trades should pay particular attention to the supply side of the labor market. Present programs may be focusing too intently on the relatively narrow issue of skill development, and not enough on the broader question of simply generating interest in the industry. Most whites, after all, do not enter the building trades through formal training. Whites, however, tend to fit better into the informal mechanisms of the industry which gradually result in a career. Efforts to help minorities compete should include ways to provide them with access to the various avenues of occupational entry, including both formal and informal training.

Questions To Be Explored

There is obviously much to be learned about minority participation in the homebuilding industry. Nor is it by any means to be inferred that the case of residential construction is in this regard reflective of the nonunion sector as a whole, or that the experience in one part of the country is the same as that in another. The issue is far from academic because information on the patterns of minority employment in nonunion construction and the reasons for those patterns will have two major practical uses: (1) it will indicate how and where this large and growing segment of an important industry can be encouraged to provide greater sources of opportunity for the disadvantaged; and (2) it will provide a more logical basis for evaluating the performance of the union sector, with significant implications for the much debated question of the relative weights of supply and demand factors in the well-documented underrepresentation of minorities in many of the skilled trades.

Research on this issue, as with most issues regarding labor in nonunion construction, must begin virtually at the beginning; even the most basic questions cannot be answered with any degree of precision. We must first ascertain the proportion of the nonunion work force that are members of various minority groups. In addition, it is almost certain that this proportion will vary by area, by industry sector, and by occupational group. Such differential patterns of minority participation must be examined in depth with an eye toward determining the factors—

historical, economic, and otherwise—which result in relatively high minority employment in one segment and relatively low employment in another.

A second important subject area is the role of nonunion contractors in the various local "plans" designed to increase minority participation in construction. In addition, as we have seen, some industry-wide training programs—those sponsored by the NAHB, for example—are specifically aimed at disadvantaged groups. The mechanics and the results of these programs have not undergone critical examination. All these programs need to be evaluated in the context of the particular characteristics of the structure of the construction industry and its labor market.

Finally, we would hope to be able to add somewhat to the developing, but still meager, literature on the minority contractor.[169] Construction is a relatively easy industry to enter, but it is also one characterized by extraordinarily high business mortality. The Glover study cited contains useful details on the background, experiences, and problems of minority employers, but is limited to a section of the country where the minority population is large and the power of unions relatively low. The situation in many northern areas may be quite different. Furthermore, there is much to be learned about the relationship between minority contractors and the various nonunion employer associations which are typically controlled by whites, and the steps these associations are taking, if any, to promote business opportunities for nonwhites.

EXTENT AND PATTERNS OF NONUNION CONSTRUCTION ACTIVITY

As noted in Chapter I, a major reason for choosing homebuilding as the focus of this study was its predominantly nonunion character. Although in many respects residential construction cannot be viewed as representative of the nonunion sector generally, this inquiry into the nature of its labor market has provided some direction for more comprehensive research. For such research to be complete, however, it cannot

---

[169] See, for example, Robert W. Glover, "The Upgrading of Minority Construction Contractors in Atlanta and Houston," Ph.D. dissertation, University of Texas, 1972; and Joseph Debro, "The Minority Builder," *Labor Law Journal*, Vol. XXI (May 1970), pp. 298-309.

simply take as given that certain segments of the industry have not been organized by the building trades unions. It must, in addition, explore the economic and historical reasons behind the varying degrees of unionization.

This study has already alluded to at least two factors which contribute to the relative weakness of unions in homebuilding. First, the industry tends to be highly dispersed and populated by large numbers of small employers. Although this is a characteristic of all construction, it is especially pronounced in the residential sector. Second, there is a tendency for firms which are engaged in building homes to specialize in that kind of activity. As such, they do not pose a major competitive threat to union standards in other sectors, and thus the unions can better tolerate open shop operations by residential contractors. Yet major questions remain. Why, for example, is homebuilding mostly nonunion in such large metropolitan areas as Buffalo, Boston, and Baltimore, whereas the unions remain strong in Chicago, Detroit, and San Francisco? Why are some residential subtrades unionized in the same localities where others are not? Why has residential construction unionism in some cities waned significantly over the past two decades? The answers are likely to revolve around a host of factors, some nationwide and some peculiar to individual areas, but these factors are at present unexplored.

The patterns of nonunion construction activity are at present no better understood outside the residential sector. The levels of nonunion strength in local building construction, highways, and major industrial and heavy building projects will also be examined in subsequent reports. In these sectors as well, it will be important to delve into the experiences of different trades, different localities, and contractors of varying size. Also of interest is the practice of "double-breasted" operations: the contractor who functions on both a union and nonunion basis. Similarly, we must pay particular attention to the number, motives, and experiences of contractors who change from union to nonunion operations.

An additional area of concern is the function of the nonunion employers' association, particularly as it relates to the labor market. It may be recalled that in Erie County, most residential subcontractors are not members of the local homebuilders' association, but there is no evidence of the extent to which this condition is also reflective of other communities or other industry sectors. At both the national and local levels, we shall inquire into the activities of such groups as the Associated Builders and Contractors, the Associated General Contractors, and the National Association of Home Builders. In addition, of course, we shall have to give special attention to the large, nationwide, nonunion firms which appear to be capturing an increasing proportion of major building work.

Another subject to be researched in this connection is the economic relationship between the union and nonunion sectors. Involved here are such questions as the influence of union wages on nonunion firms, the movements of both workers and contractors from one sector to the other, and the business dealings between unionized contractors and nonunion subcontractors (and vice versa). As noted above, these relationships are apt to be stronger in nonresidential construction than in homebuilding.

Finally, and perhaps of principal interest to some, there is the unions' response to the growing competition from open shop contractors. One such response involves the negotiation of project agreements, in which a special multicraft set of guidelines is promulgated for a single undertaking. Normally, these agreements include concessions by the unions, sometimes involving the freezing of existing wage rates for the duration of the project or waiving certain rules or benefits contained in local labor-management contracts. In some areas these concessions have not been limited to the project agreement; the unions may informally neglect to enforce onerous provisions in order to allow their contractors to compete more effectively with nonunion firms. In other cases, the union response has taken the form of violence against open shop contractors: recently the National Labor Relations Board issued complaints against four national and eleven local construction unions based on evidence that violence was utilized against open shop contractors and employees. This is an indication of the seriousness of the union-open shop conflict. These are some of the questions which we hope to answer in our forthcoming study of the entire open shop segment of the construction industry.

Index

LABOR RELATIONS AND PUBLIC POLICY SERIES

9. *NLRB and Management Decision Making*, by Robert A. Swift. 1974. $6.95

8. *NLRB Regulation of Election Conduct*, by Robert E. Williams, Peter A. Janus, and Kenneth C. Huhn. 1974. $8.50

7. *Opening the Skilled Construction Trades to Blacks: A Study of the Washington and Indianapolis Plans for Minority Employment*, by Richard L. Rowan and Lester Rubin. 1972. $5.95

6. *Welfare and Strikes: The Use of Public Funds to Support Strikers*, by Armand J. Thieblot, Jr., and Ronald M. Cowin. 1972. $6.95

5. *Collective Bargaining: Survival in the '70's?* Conference Proceedings, edited by Richard L. Rowan. 1972. $8.50

4. *The NLRB and Secondary Boycotts*, by Ralph M. Dereshinsky. 1972. $5.95

3. *The NLRB and the Appropriate Bargaining Unit*, by John E. Abodeely. 1971. $5.95

OTHER COLLECTIVE BARGAINING STUDIES

William N. Chernish, *Coalition Bargaining.* Major Study No. 45. 1969. $7.95

Herbert R. Northrup and Gordon R. Storholm, *Restrictive Labor Practices in the Supermarket Industry.* Major Study No. 44. 1967. $7.50

Michael H. Moskow, *Teachers and Unions: The Applicability of Collective Bargaining to Public Education.* Major Study No. 42. 1966. $8.50

INDUSTRY STUDIES

F. Marion Fletcher, *Market Restraints in the Retail Drug Industry.* Major Study No. 43. 1967. $10.00

Robert W. Kirk, *The Carpet Industry: Present Status and Future Prospects.* Miscellaneous Report Series No. 17. 1970. $5.95

George M. Parks, *The Economics of Carpeting and Resilient Flooring: An Evaluation and Comparison.* Major Study No. 41. 1966. $2.95